Life
and Flowers

Life
and Flowers

Life and Flowers

By Maurice Maeterlinck

linck ✤ ✤ *Translated by* Alexander Teixeira de Mattos

Fredonia Books
Amsterdam, The Netherlands

Life and Flowers

by
Maurice Maeterlinck

ISBN: 1-58963-690-2

Fredonia Books
Amsterdam, The Netherlands
http://www.fredoniabooks.com

NOTE

Of the essays forming this volume, some two or three are now published, in English, for the first time. The remainder have appeared at different times in newspapers and magazines, including the Fortnightly Review, Harper's Magazine, *the* Atlantic Monthly, *the* New York Critic, *the* International Quarterly, *the* Daily Mail *and others. The thanks of the author and translator are due to the proprietors and editors of these periodicals for their leave to republish in the present volume.*

CONTENTS

	PAGE
THE MEASURE OF THE HOURS .	3
IMMORTALITY .	17
THE GODS OF WAR	51
OUR SOCIAL DUTY .	65
OUR ANXIOUS MORALITY .	85
ROME	127
THE PSYCHOLOGY OF ACCIDENT	145
IN PRAISE OF THE FIST	163
THE FORGIVENESS OF INJURIES .	175
CONCERNING "KING LEAR"	189
THE INTELLIGENCE OF THE FLOWERS	205
PERFUMES .	301

THE MEASURE OF THE HOURS

A

THE MEASURE OF THE HOURS

I

SUMMER is the season of happiness. When, among the trees, in the mountains or by the sea, the fair hours of the year, the hours for which we have waited and hoped since the depths of winter, the hours which at last open to us the golden gates of leisure, return for our delight, let us learn to enjoy them fully, continuously, voluptuously. Let us have for these privileged hours a nobler measure than that into which we pour the ordinary hours. Let us gather their dazzling minutes in unaccustomed urns, glorious, transparent and made of the very light which they are to contain, even as we serve a costly wine not in the common glass of the daily table, but in the purest cup of crystal and silver locked in the sideboard of the banqueting-room.

Life and Flowers

The measuring of time! We are so constructed that we cannot be made conscious
of time and impressed with its joys or sorrows
unless we count and weigh it, like an invisible currency. It takes shape, acquires its
substance and its value only in complicated
forms of apparatus which we have contrived
in order to render it apparent; and, having no
existence in itself, it borrows the taste, the
perfume and the shape of the instrument that
rules it. For this reason, the minutes ticked
off by our little watches wear a different
aspect from those prolonged by the great
hand of the belfry or cathedral-clock. It
behoves us, therefore, not to be indifferent
to the birth of the hours. Even as we have
glasses whose shape, shade and brilliancy
vary according as they are called upon to
carry to our lips light claret or rich bur-
gundy, cool hock or heavy port, or the
gladness of champagne, why should not our
minutes be numbered in ways appropriate
to their melancholy, their inertness or their
joy? It is fitting, for instance, that our

The Measure of the Hours

working months and our winter days, days of bustle, business, hurry and restlessness, should be strictly, methodically, harshly divided and registered by the metal wheels and hands and the enamelled faces of our chimney-clocks, our electric or pneumatic dial-plates or our small pocket-watches. Here, majestic time, the master of gods and men, the immense human form of eternity, is no more than a stubborn insect gnawing mechanically at a life devoid of horizon, sky or rest. At most, at the warning moment that precedes the stroke, during the too-short evening snatched, under the lamp, from the cares of hunger or vanity, will the great copper pendulum of the Dutch or Norman clock be allowed to make slower and more impressive the seconds that go before the steps of grave night advancing.

3

On the other hand, for our hours no longer indifferent, but really sombre, for our hours of discouragement, of self-denial, of sickness and pain, for the dead minutes of our life, let us regret the time-honoured, dejected and silent hour-glass of our ancestors. It is to-day no more than an inactive symbol on our tombstones or the funeral hangings of our churches, except where, pitifully fallen, it may still be found presiding, in some country kitchen, over the fastidious cooking of our boiled eggs. It no longer continues as an instrument of time, though it still figures, in company with the scythe, on its antiquated blazon. And yet it had its merits and its reasons for existence. In the dull, sad days of human thought, in the cloisters built around the abode of the dead, in the convents that opened their doors and windows only to the wavering glimmers of another world, more awful than our own, the sand-glass was, for the hours stripped of their joys, their smiles, their happy surprises and their ornaments, a measure whose place no other could have filled as gracefully. It did not state time with precision ; it stifled it in powdery particles. It was made for counting one by one the sands of prayer and waiting, of terror and weariness. The minutes sped by in dust, isolated from the circumambient life

The Measure of the Hours

of the sky, the garden and space, secluded
in their glass phial even as the monk was
secluded in his cell, marking, naming no
hour, burying them all in the funeral sand,
while the unoccupied thoughts that watched
over their dumb and incessant fall passed
away with them to be added to the ashes
of the dead.

4

Between the glorious banks of flaming
summer, it seems best to enjoy the glowing
succession of the hours in the order in which
they are marked by the orb itself that
showers them upon our leisure. In these
wider, more open, more lingering days, I
believe and trust only in the great divisions
of light which the sun names to me with the
warm shadow of its rays on the marble dial
which there, in the garden, beside the lake,
reflects and records in silence, as though it
were doing an insignificant thing, the course
of our worlds through planetary space. By
this immediate, this only authentic tran-
scription of the wishes of time which
directs the stars, our poor human hour,

which rules our meals and all the little
actions of our little lives, acquires a nobility,
a direct and urgent fragrance of infinity
that render vaster and more health-giving
the dazzling, dewy mornings and almost
motionless afternoons of the fair and immaculate summer.

Unfortunately, the sun-dial, which alone
knew how to follow with dignity the grave
and luminous march of the spotless hours,
is becoming rare and is disappearing from our
gardens. It is hardly anywhere to be found
save in the main court, on the stone terraces,
in the mall, among the quincunxes of some
old town, some old castle, some ancient palace,
where its gilt figures, its face and style are
wearing away under the hand of the very
god whose worship they should perpetuate.
Nevertheless, Provence and some of the
Italian market-towns have remained faithful
to the celestial clock. Here we often see displayed, on the sunny gable of the brightest
of old, dilapidated country-houses, the fres-
coed circle over which the sunbeams carefully
measure their fairy progress. And mottoes,
profound or artless, but always significant,

because of the place which they fill and the part which they play in a vast life, strive to blend the human soul with incomprehensible phenomena. "*L'heure de la justice ne sonne pas aux cadrans de ce monde:* the hour of justice does not strike on the dials of this world," says the inscription on the sun-dial of the church at Tourette-sur-Loup, that extraordinary, that almost African little village, near to where I live, which, amid the crumbling rocks and clambering aloes and fig-trees, resembles a miniature Toledo reduced to a skeleton by the sun. Another radiant clock-face proudly proclaims "*A lumine motus*" as its motto: "I am moved by the light." "Amyddst yᵉ flovvres, I tell yᵉ houres," says an old marble dial in an old garden. But one of the prettiest legends, surely, is that which Hazlitt discovered one day near Venice: "*Horas non numero nisi serenas.*"

"'I count only the hours that are serene,'" he adds. "What a bland and care-dispelling feeling! How the shadows seem to fade on the dial-plate as the sky lours and time presents only a blank, unless as its progress is marked by what is joyous and all that is not happy descends into oblivion. What a fine lesson is conveyed to the mind to take no note of time but by its benefits, to watch only for the smiles and neglect the frowns of fate, to compose our lives of bright and gentle moments, turning always to the sunny side of things and letting the rest slip from our imaginations, unheeded or forgotten!"

5

The clock, the hour-glass, the vanished clepsydra give abstract hours, without face or form. They are the instruments of the anæmic time of our indoor rooms, of time enslaved and captive; but the sun-dial reveals to us the real, throbbing shadow of the wing of the great god that hovers in the sky. Around the marble disk which adorns the terrace or the crossing of two wide avenues and which harmonizes so well with the majestic staircases and spreading balustrades or with the green walls of the thick quickset hedges, we enjoy the fleeting but undeniable presence of the beamy hours. He who has learnt to descry them in space will see them turn by turn touching earth and leaning over the mysterious altar to offer a sacrifice to the god whom man honours, but cannot know. He will see them advancing in diverse and changing garments, crowned with fruit, flowers or dew: first, the as yet diaphanous and hardly visible hours of the dawn; next, their sisters of noon, ardent, cruel, resplendent, almost implacable; and, finally, the last hours of twilight, slow and sumptuous, delayed in their progress towards approaching night by the purpling shadow of the trees.

6

The sun-dial alone is worthy to measure the splendour of the months of green and gold. Like profound happiness, it speaks no word. Time marches over it in silence, as it passes in silence over the spheres of space; but the church of the neighbouring village lends it at moments its bronze voice; and nothing is so harmonious as the sound of the bell that strikes a chord with the

dumb gesture of its shadow marking noon
amid the sea of blue. The sun-dial gives a
centre and successive names to scattered and
nameless joys. All the poetry, all the
delights of the country-side, all the mysteries of the firmament, all the confused
thoughts of the tall trees that guard like a
sacred treasure the coolness which night has
entrusted to their care, all the blissful intensity of the corn-fields, plains and hills
abandoned without defence to the devouring
magnificence of the sunlight, all the indolence
of the brook flowing between its gentle
banks, the drowsiness of the pond covered
with drops of sweat formed by the duck-
weed, the satisfaction of the house that
opens, in its white front, windows greedy
to draw in the horizon, the scent of the
flowers hastening to finish a day of scorch-
ing beauty, the birds singing in the order
of the hours to weave garlands of gladness
for them in the sky: all these, together
with thousands of things and thousands of
lives that escape our sight, meet and take
stock of their continuance around this
mirror of time on which the sun, which

The Measure of the Hours

is but one of the wheels of the huge
machine that vainly subdivides eternity,
marks with a kindly ray the daily journey
which the earth, with all that it carries,
performs on the road of the stars.

IMMORTALITY

IMMORTALITY

I

In the new era whereupon we are entering and wherein the religions no longer reply to the great questions of mankind, one of the problems on which we cross-examine ourselves most anxiously is that of the life beyond the tomb. Do all things end at death? Is there an imaginable after-life? Whither do we go and what becomes of us? What awaits us on the other side of the frail illusion which we call existence? At the moment when our heart stops beating, does matter triumph, or mind; does eternal light begin, or endless darkness?

Like all that exists, we are imperishable. We cannot conceive that anything should be lost in the universe. By the side of infinity, it is impossible to imagine a state of nothingness into which an atom of matter can fall and be annihilated. All that is will be eternally; all is; and there is nothing that is not. Otherwise, we should be driven to believe that our brain has nothing in common with the universe which it strives to conceive. We should even have to say that it works in the reverse direction to the universe, which is hardly probable, since it is, after all, perhaps but a sort of reflection of the universe.

That which appears to perish or, at least, to disappear and follow upon itself is the form and fashion under which we see imperishable matter; but we do not know with what realities these appearances correspond. They are the texture of the bandage which is laid upon our eyes and which gives them, under the pressure that blinds them, all the images of our life. Remove that bandage: what remains? Do we enter into the reality that undoubtedly exists beyond? Or do the appearances themselves cease to exist for us?

Immortality

2

That the state of nothingness is impossible;
that, after our death, all subsists in itself and
nothing perishes : these are things that hardly
interest us.   The only point that touches us,
in this eternal persistence, is the fate of that
small part of our life which used to perceive
phenomena during our existence.   We call it
our consciousness or our ego.   This ego, as
we conceive it when we reflect upon the
consequences of its destruction, this ego is
neither our mind nor our body, since we
recognize that both are waves that flow
away and are renewed incessantly.   Is it an
immovable point, which could not be form or
substance, for these are always in evolution,
nor life, which is the cause or effect of form
and substance ?   In truth, it is impossible for
us to apprehend or define it, to tell where it
dwells.   When we try to go back to its last
source, we find hardly more than a succession of memories, a series of ideas, confused,
for that matter, and unsettled, attached
to the one instinct of living : a series of
habits of our sensibility and of conscious or

unconscious reactions against the surrounding phenomena. When all is said, the most steadfast point of that nebula is our memory, which seems, on the other hand, to be a somewhat external, a somewhat accessory faculty and, in any case, one of the frailest faculties of our brain, one of those which disappear the most promptly at the least disturbance of our health.

3

It matters not; that uncertain, indiscernible, fleeting and precarious ego is so much the centre of our being, interests us so exclusively that every reality of our life disappears before this phantom. It is a matter of utter indifference to us that thoughout eternity our body or its substance should know every joy and every glory, undergo the most splendid and delightful transformations, become flower, perfume, beauty, light, air, star; it is likewise indifferent to us that our intellect should expand until it mixes with the life of the worlds, understands and governs it. Our

Immortality

instinct is persuaded that all this will not affect us, will give us no pleasure, will not happen to ourselves, unless that memory of a few almost always insignificant facts accompany us and witness those unimaginable joys. I care not if the loftiest, the freest, the fairest portions of my mind be eternally living and radiant in the supreme gladnesses: they are no longer mine; I do not know them. Death has cut the network of nerves or memories that connected them with I know not what centre wherein lies the sensitive point which I feel to be all myself. They are now set loose, floating in space and time, and their fate is as unknown to me as that of the farthest stars. Anything that occurs exists for me only upon condition that I be able to recall it within that mysterious being which is I know not where and precisely nowhere, which I turn like a mirror about this world whose phenomena take shape only in so far as they are reflected in it.

Life and Flowers

4

Thus our longing for immortality destroys itself while expressing itself, since it is on one of the accessory and most transient parts of our whole life that we base all the interest of our after-life. It seems to us that, if our existence be not continued with the greater part of its drawbacks, of the pettinesses and blemishes that characterize it, nothing will distinguish it from that of other beings; that it will become a drop of ignorance in the ocean of the unknown; and that, thenceforth, all that may ensue will no longer concern us.

"What immortality can one promise to men who almost necessarily conceive it in this guise? How can we help it?" asks a puerile, but profound instinct. Any immortality that does not drag with it through eternity, like the fetters of the convict that we were, the strange consciousness formed during a few years of movement, any immortality that does not bear that indelible mark of our identity is for us as though it were not. Most of the religions have understood this and have taken account of that instinct which desires and at the same time destroys the after-life. It is thus that the Catholic Church, going back to the most primitive hopes, promises us not only the integral preservation of our earthly ego, but even the resurrection of our own flesh.

There lies the crux of the riddle. When we demand that this small consciousness, that this sense of a special ego—almost childish and, in any case, extraordinarily limited : probably an infirmity of our actual intelligence—should accompany us into the infinity of time in order that we may understand and enjoy it, are we not wishing to perceive an object with the aid of an organ that is not intended to perceive it? Are we not asking that our hand should discover the light or that our eye should take in perfumes? Are we not, on the other hand, acting like a sick man who, in order to recognize himself, to be quite sure that he is himself, should think it necessary to continue his sickness in his health and in the boundless sequence of his days? The comparison, for that matter, is more accurate than is the habit of comparisons. Picture a blind man who is also paralyzed and deaf. He has been in this condition from his birth and has just attained his thirtieth year. What can the hours have embroidered on the imageless web of this poor life'? The unhappy man must have gathered in the depths of his memory, for lack of other recollections, a few wretched sensations of heat and cold, of weariness and rest, of more or less keen physical sufferings, of hunger and thirst. It is probable that all human joys, all our ideal hopes and dreams of paradise will be reduced for him to the confused sense of well-being that follows the allaying of a pain. This, then, is the only possible equipment of that consciousness and that ego. The intellect, having never been invoked from without, will sleep soundly, knowing nothing of itself. Nevertheless, the poor wretch will have his little life to which he will cling by bonds as narrow and eager as though he were the happiest of men. He will dread death; and the idea of entering into eternity without carrying with him the emotions and memories of his dark and silent sick-bed will plunge him into the same despair into which we are plunged by the thought of abandoning a life of glory, light and love for the icy darkness of the tomb.

5

Let us now suppose that a miracle suddenly quicken his eyes and ears and reveal to him, through the open window at the head of his bed, the dawn rising over the plain, the song of the birds in the trees, the murmuring of the wind in the leaves and of the water against its banks, the ringing of human voices among the morning hills. Let us suppose also that the same miracle, completing its work, restore the use of his limbs. He rises, stretches his arms to that prodigy which as yet for him possesses neither reality nor name: the light! He opens the door, staggers out amidst the effulgence; and his whole body dissolves in all these marvels. He enters upon an ineffable life, upon a sky whereof no dream could have given him a foretaste; and, by

a freak which is readily admissible in this
sort of cure, health, introducing him to this
inconceivable and unintelligible existence,
wipes out in him all memory of days past.

What will be the state of that ego, of
that central focus, the receptacle of all our
sensations, the spot in which converges all
that belongs in its own right to our life, the
supreme point, the "egotic" point of our
being, if I may venture to coin a word?
Memory being abolished, will that ego re-
cover within itself a few traces of the man
that was?   A new force, the intellect,
awaking and suddenly displaying an unprecedented activity, what relation will that
intellect keep up with the inert, dull germ
whence it has sprung?   At what corner of
his past will the man clutch to continue his
identity?   And yet will there not survive
within him some sense or instinct, independent of the memory, the intellect and
I know not what other faculties, that will
make him recognize that it is indeed in
him that the liberating miracle has been
wrought, that it is indeed his life and not his
neighbour's, transformed, irrecognizable, but

substantially the same, that has issued from
the silence and the darkness to prolong itself
in harmony and light? Can we picture the
disarray, the flux and reflux of that bewildered consciousness? Have we any idea
in what manner the ego of yesterday will
unite with the ego of to-day and how the
" egotic" point, the sensitive point of the
personality, the only point which we are
anxious to preserve intact, will bear itself
in that delirium and that upheaval?

Let us first endeavour to reply with sufficient preciseness to this question which
comes within the scope of our actual and
visible life; and, if we are unable to do this,
how can we hope to solve the other problem
that presents itself before every man at the
moment of his death?

6

This sensitive point, in which the whole
problem is summed up, for it is the only
one in question and, except in so far as it is
concerned, immortality is certain ; this mysterious point, to which, in the presence of

death, we attach so high a value, we lose, strange to say, at any moment in life without feeling the least anxiety. Not only is it destroyed nightly in our sleep, but even in waking it is at the mercy of a host of accidents. A wound, a shock, an indisposition, a few glasses of alcohol, a little opium, a little smoke are enough to obliterate it. Even when nothing impairs it, it is not constantly perceptible. We often need an effort, in turning back upon ourselves to recapture it, to become aware that such or such an event is occurring to us. At the least distraction, a happiness passes beside us without touching us, without yielding up to us the pleasure which it contains. One would say that the functions of the organ by which we taste life and bring it home to ourselves are intermittent, often interrupted or suspended and that the presence of our ego, except in pain, is but a rapid and perpetual sequence of departures and returns. What reassures us is that we believe ourselves sure to find it intact on awaking, after the wound, the shock or the distraction, whereas we are persuaded, so

fragile do we feel it to be, that it is bound
to disappear for ever in the terrible concus-
sion that separates life from death.

7

One foremost truth, pending others which
the future will no doubt reveal, is that,
in these questions of life and death, our
imagination has remained very childish.
Almost every elsewhere, it precedes reason;
but here it still loiters over the games of
the earliest ages. It surrounds itself with
the dreams and the barbarous longings
wherewith it lulled the hopes and fears of
cave-dwelling man. It asks for things that
are impossible, because they are too small.
It claims privileges which, if obtained, were
more to be dreaded than the most enormous disasters with which nihility threatens
us. Can we think without shuddering of an
eternity contained wholly within our infinitesimal actual consciousness? And behold
how, in all this, we obey the illogical whims
of what used to be called *la folle du logis!*
Which of us, if he went to sleep to-night

in the scientific certainty of awaking in a
hundred years as he is to-day, with his
body intact, even on condition that he lost
all memory of his previous life—would
those memories not be useless?—which of
us would not welcome that secular sleep
with the same confidence as the brief, gentle
slumbers of his every night? Far from
dreading it, would not many hasten to
make the trial with eager curiosity? Should
we not see numbers of men assail the dis-
penser of the fairy sleep with their prayers
and implore as a favour what they would
deem a miraculous prolongation of their
life? And yet, during that sleep, how
much would remain and how much of
themselves would they find again on
awaking? What link, at the moment
when they closed their eyes, would connect
them with the being that was to awake
without memories, unknown, in a new
world? Nevertheless, their consent and
all their hopes at the beginning of that
long night would depend upon that non-
existing link. There is, in fact, between
real death and this sleep only the difference

of that awakening deferred for a century, an awakening here as alien to him who had gone to sleep as the birth would be of a posthumous child.

8

On the other hand, what answer do we make to the question when it has to do not with us, but with the things that breathe with us on earth? Are we concerned, for instance, about the after-life of the animals? The most faithful, affectionate and intelligent dog, once dead, becomes but a repulsive carcass, which we hasten to get rid of. It does not even seem possible to us to ask ourselves if any part of the already spiritual life which we loved in him subsist elsewhere in our memory, or if there be another world for dogs. It would appear rather ridiculous to us that time and space should preserve preciously, for all eternity, among the stars and the boundless mansions of the sky, the soul of a poor beast, made up of five or six touching, but very unsophisticated habits and of the longing to eat and drink, to sleep warm and to greet his kind in the manner which we know. Besides, what could remain of that soul, composed entirely of the few needs of a rudimentary body, when that body has ceased to exist? Yet by what right do we imagine between ourselves and the animal an abyss that does not exist even between the mineral and the vegetable, the vegetable and the animal? This right to believe ourselves so far, so different from all that lives upon earth, this pretension to place ourselves in a category and a kingdom to which the very gods whom we have created would not always have access: these are what we must first of all examine.

9

It would be impossible to set forth all the paralogisms of our imagination on the point which we are discussing. Thus, we are pretty easily resigned to the dissolution of our body in the grave. We are not at all anxious that it should accompany us in the infinity of time. Upon reflection, we should even be vexed were it to escort us with its inevitable drawbacks, its faults, its blemishes, and its absurdities. What we intend to take with us is our soul. But what shall we answer to one who asks us if it be possible to conceive that this soul is anything more than the sum total of our intellectual and moral faculties, added, if you will—to make full measure—to all those which fall within the jurisdiction of our instinct, our unconsciousness, our subconsciousness? Now, when, at the approach of old age, we see these same faculties become impaired, in either ourselves or others, we do not distress ourselves, we do not despair any more than we distress ourselves or despair when we behold the slow decline of our physical strength. We keep intact our dim hope of an after-life. It seems to us quite natural that the state of one set of faculties should depend upon the state of the others. Even when the former are completely destroyed in a being whom we love, we do not consider that we have lost him or that he has lost his ego, his moral personality, of which, however, nothing remains. We should not mourn his loss, we should not think that he was no more, if death preserved those

faculties in their state of annihilation. But,
if we do not attach a capital importance to
the dissolution of our body in the tomb, or
to the dissolution of our intellectual faculties
during life, what is it that we ask death to
spare and of what unrealizable dream do we
demand the realization?

10

In truth, we cannot, at least for the
moment, imagine an acceptable answer to
the question of immortality. Why be
astonished? Here stands my lamp on my
table. It contains no mystery; it is the
oldest, the best known and the most
familiar object in the house. I see in it
oil, a wick, a glass chimney; and all of this
forms light. The riddle begins only when
I ask myself what this light is, whence it
comes when I call it, where it goes when
I extinguish it. Then, suddenly, around
this small object which I can lift, take to
pieces and which might have been fashioned
by my hands, the riddle becomes unfathom-
able. Gather round my table all the men
that live upon this earth: not one will be

able to tell us what this little flame is which
I cause to take birth or to die at my
pleasure. And, should one of them venture
upon one of those definitions known as
scientific, every word of the definition will
multiply the unknown and, on every side,
open unexpected doors into endless night.
If we know nothing of the essence, the
destiny, the life of a gleam of familiar light
of which all the elements were created by
ourselves, of which the source, the proxi-
mate causes and the effects are contained
within a china bowl, how can we hope to
penetrate the mystery of a life of which the
simplest elements are situated at millions of
years, at thousands of millions of leagues
from our intelligence in time and space?

II

Since humanity began to exist, it has not
advanced a single step on the road of the
mystery which we are contemplating. No
question which we ask ourselves on the subject touches, on any side, the sphere in
which our intelligence is formed and moves.

Life and Flowers

There is perhaps no relation possible or imaginable between the organ that puts the question and the reality that ought to reply to it. The most active and searching enquiries of late years have taught us nothing. Learned and conscientious psychical societies, notably in England, have got together an imposing collection of irrefutable facts which prove that the life of the spiritual or nervous being can continue for a certain time after the death of the material being. No sincere mind now dreams of denying the possibility of these facts supported by documentary and other evidence as conclusive as that which serves as a basis for our firmest scientific convictions. But all this merely removes by a few lines, by a few hours the beginning of the mystery. If the ghost of a person whom I love, clearly recognizable and apparently so much alive that I speak to it, enter my room to-night at the very minute when life is quitting the body at a thousand miles from the spot where I am, that is, no doubt, very strange, even as everything is strange in a world of which we do not understand the first word; but it shows at most that the soul, the spirit, the breath, the nervous and indiscernible force of the subtlest part of our matter can disengage itself from that matter and survive it for an instant, even as the flame of a lamp which we extinguish sometimes becomes detached from the wick and hovers for a moment in the darkness. Certainly, the phenomenon is an astonishing one; but, given the nature of that spiritual force, we ought to be much more astonished that it is not produced more frequently and at our pleasure, in the fulness of life. In any case, it throws no light upon the question. Never has a single one of those phantasms appeared to have the least consciousness of a new life, of a supraterrestrial life, a life different from that just abandoned by the body whence it emanated. On the contrary, the spiritual life of all of them, at that moment when it ought to be pure, since it is rid of matter, seems greatly inferior to what it was when enveloped in matter. Most of them, in a sort of somnambulistic dulness, pursue mechanically the most insignificant of their accustomed preoccupations. One looks for its hat, which it has left on a chair or table ; another is troubled about a small debt or anxious to know the time. And all of them, a little later, at the moment when the real after-life ought to begin, evaporate and disappear for ever. I agree that this proves nothing either for or against the possibility of the after-life. We do not know whether these brief apparitions be the first glimmers of a new or the last of the present existence. Perhaps the dead, for want of a better, thus use and turn to account the last bond that links them and makes them still perceptible to our senses. Perhaps, afterwards, they continue to live around us, but fail, despite all their efforts, to make themselves recognized or to give us an idea of their presence, because we have not the organ needed to perceive them, even as all our efforts would fail to give a man blind from birth the least notion of light or colour. In any case, it is certain that the investigations and the labours of that new science of the " Borderland," as the English call it, have left the problem exactly where it has been since the beginning of human consciousness.

Immortality

In the invincible ignorance, then, in which we are, our imagination has the choice of our eternal destinies. Now, when we examine the different possibilities, we are compelled to admit that the most beautiful are not the least probable. A first hypothesis, to be put aside off-hand, without discussion, as we have seen, is that of absolute annihilation. A second hypothesis, eagerly cherished by our blind instincts, promises us the more or less integral preservation, through the infinity of time, of our consciousness or our actual ego. We have also studied this hypothesis, which is a little more plausible than the first, but at bottom so narrow, so naive and puerile that, whether for men or for plants and animals, one sees scarcely a means of finding a reasonable place for it in boundless space and infinite time. Let us add that, of all our possible destinies, it would be the only one to be really dreaded and that annihilation pure and simple would be a thousand times preferable.

There remains the double hypothesis of

an after-life without consciousness, or with an enlarged and transformed consciousness, of which that which we possess to-day can give us no idea, which it prevents us rather from conceiving, even as our imperfect eye prevents us from conceiving any other light than that which goes from infra-red to ultra-violet, whereas it is certain that those probably prodigious lights would dazzle on every side, in the darkest night, a pupil shaped differently from ours.

Although double at the first view, the hypothesis is soon brought back to the simple question of consciousness. To say, for instance, as we are tempted to do, that an after-life without consciousness is equivalent to annihilation is to settle *a priori* and without reflection that problem of consciousness which is the chief and most obscure of all the problems that interest us. It is, as all the metaphysicians have proclaimed, the most difficult that exists, considering that the object of our knowledge is the very thing that is striving to know. What, then, can that mirror ever facing itself do, save reflect itself indefinitely and

to no purpose ? Nevertheless, in that reflection incapable of emerging from the
multiplication of itself sleeps the only ray
that is able to throw light upon all the rest.
What is to be done? There is no other
means of escaping from one's consciousness
than to deny it, to look upon it as an
organic disease of the terrestrial intelligence,
a disease which we must endeavour to cure
by an act which must appear to us an
act of violent and wilful madness, but
which, on the other side of our seeming,
is probably an act of sanity.

13

But escape is impossible; and we return fatally to prowl around our consciousness based upon our memory, the
most precarious of all our faculties. It
being evident, we say, that nothing can
perish, we must needs have lived before our
present life. But, as we are unable to
connect our previous existence with our
actual life, this certainty is as indifferent to
us, passes as far from us as all the certainties

of our later life. And here we have, before life as after death, the appearance of the mnemonic ego, concerning which it behoves us once more to ask ourselves if what it does during the few days of its activity is really important enough thus to decide, by reference to itself alone, the problem of immortality. From the fact that we enjoy our ego under so exclusive, special, imperfect, fragile and ephemeral a form does it follow that there is no other mode of consciousness, no other means of enjoying life? A nation of men born blind, to return to the comparison which becomes essential, because it best sums up our situation in the midst of the darkness of the worlds, a nation of men blind from their birth, to whom a solitary traveller should reveal the joys of the light, would deny not only that the latter was possible, but even imaginable. As for ourselves, is it not very nearly certain that we lack here below, among a thousand other senses, a sense superior to that of our mnemonic consciousness, in order to have a fuller and surer enjoyment of our ego? May it not be said that we sometimes catch

Immortality

obscure traces or feeble desires of that budding or atrophied sense, oppressed in any
case and almost suppressed by the rule of
our terrestrial life, which centralizes all the
evolutions of our existence upon the same
sensitive point? Are there not certain confused moments in which, however ruthlessly,
however scientifically we may allow for
egoism pursued to its most remote and
secret sources, there remains in us something absolutely disinterested that takes
pleasure in the happiness of others? Is it
not also possible that the aimless joys of art,
the calm and deep satisfaction into which
we are plunged by the contemplation of a
beautiful statue, of a perfect building, which
does not belong to us, which we shall never
see again, which arouses no sensual desire,
which can be of no service to us: is it not
possible that this satisfaction may be the pale
glimmer of a different consciousness that
filters through a cranny of our mnemonic
consciousness? If we are unable to imagine
that different consciousness, that is no reason
for denying it. I even believe that it would
be wiser to assert that this would be a reason

for admitting it. All our life would be spent in the midst of things which we could never have imagined, if our senses, instead of being given to us all together, had been granted to us one by one and from year to year. For that matter, one of these senses, the sense of generation, which awakens only at the approach of puberty, shows us that the discovery of an unexpected world, the displacement of all the axes of our life depends upon an accident of our organism. During childhood, we did not suspect the existence of a whole world of passions, of love's frenzies and sorrows which excite "grown-up people." If some garbled echo of those sounds, by chance, happened to reach our innocent and curious ears, we did not succeed in understanding what manner of fury or madness was thus seizing hold of our elders, and we promised ourselves, when the time came, to be more sensible, until the day when love unexpectedly appearing disturbed the centre of gravity of all our feelings and of most of our ideas. We see, therefore, that to imagine or not to imagine depends upon so little that we have

44

no right to doubt the possibility of that
which we cannot conceive.

14

What keeps and will long still keep us
from enjoying the treasures of the universe
is the hereditary resignation with which we
tarry in the gloomy prison of our senses.
Our imagination, as we lead it to-day, ac-
commodates itself too readily to that
captivity. True, it is the slave of those
senses which alone feed it. But it does
not sufficiently cultivate within itself the
intuitions and presentiments which tell it that
it is kept absurdly captive and that it should
seek outlets even beyond the most resplendent and infinite circles which it pictures to
itself. It is important that our imagination
should say to itself, with ever-increasing
seriousness, that the real world begins
thousands of millions of leagues beyond its
most ambitious and daring dreams. Never
was it entitled, nay, bound to be more
madly reckless than now. All that it
succeeds in building and multiplying in the

Life and Flowers

most enormous space and time that it is
capable of conceiving is as nothing compared
with what is. Already the smallest revelations
of science in our humble daily life teach it
that, even in this modest environment, it
is unable to cope with reality, that it is being
constantly overwhelmed, bewildered, dazzled
by all the unexpected that lies hidden in a
stone, a grain of salt, a glass of water, a
plant, an insect. It is already something to
be convinced of this, for that places us in a
state of mind that watches every occasion to
break through the magic circle of our
blindness; it persuades us also that we
cannot hope to find decisive truths within
this circle, that they all lie beyond it.
Man, to maintain his sense of proportion,
has a need to tell himself at every moment
that, were he suddenly placed amid the
realities of the universe, he would be exactly
comparable with an ant which, knowing
only the narrow pathways, the tiny holes,
the approaches and horizons of its ant-
heap, should suddenly find itself floating
on a straw in the midst of the Atlantic.
Pending the time when we shall have left a

prison which prevents us from coming into touch with the realities beyond our imagination, we stand a much greater chance of lighting upon a fragment of truth by imagining the most unimaginable things than by striving to lead the dreams of that imagination, through the midst of eternity, between the dikes of logic and of actual possibilities. Let us therefore try, whenever a new dream presents itself, to snatch from before our eyes the bandage of our earthly life. Let us say to ourselves that, among all the possibilities which the universe still hides from us, one of the easiest to realize, one of the most probable, the least ambitious and the least disconcerting is certainly the possibility of enjoying an existence much more spacious, lofty, perfect, durable and secure than that which is offered to us by our actual consciousness. Admitting this possibility—and there are few as probable—the problem of our immortality is, in principle, solved. It now becomes a question of grasping or foreseeing its ways and, amid the circumstances that interest us the most, of knowing what part of our intellectual and moral

Life and Flowers

acquirements will pass into our eternal and
universal life. This is not the work of
to-day or to-morrow; but it would need no
incredible miracle to make it the work of
some other day. . . .

THE GODS OF WAR

THE GODS OF WAR

I

WAR ever offers a magnificent theme for the meditations of men and one that is incessantly renewed. It remains certain, alas, that most of our efforts and inventions are always converging towards it, making of it a sort of diabolical mirror in which the progress of our civilization is reflected upside down.

I propose to-day to look at it from only one point of view, in order once again to establish the fact that the more we triumph over the unknown forces the more we yield to them. No sooner have we perceived in the obscurity or the apparent sleep of nature a new glimmer, a new source of energy than we often become its victims and nearly always its slaves. It is as though, thinking to free ourselves, we freed formidable enemies. True it is that, in the long run, those enemies end by allowing themselves to be led and render us services wherewith we could no longer dispense. But hardly has one of them made its submission before, in the very act of passing under the yoke, it places us on the track of an infinitely more dangerous adversary; and thus our fate becomes more and more glorious and more and more uncertain. Moreover, among these adversaries are some that seem quite indomitable. But perhaps they remain refractory only because they know better than the others how to appeal to those evil instincts of our heart which delay by many centuries the conquests of our intelligence.

2

This is notably the case with the majority of the inventions that relate to war. We have seen this in recent monstrous conflicts. For the first time since the beginning of history, entirely new forces, mature at last, have emerged from the darkness of a long period

of experiment and probation and come to
take the place of men on the battlefield.
Until the late wars, these forces still hung
back, held themselves aloof and acted only
from afar. They were reluctant to assert
themselves; and there was still some connection between their mysterious action and the
work of our own hands. The range of the
rifle was not greater than that of our eye;
and the destructive energy of the most murderous gun, of the most formidable explosive
still preserved human proportions. To-day,
we are overwhelmed, we have definitely abdi-
cated, our reign is ended; and behold us, as
so many grains of sand, at the mercy of
the monstrous and enigmatic powers whose
aid we have dared to invoke.

3

It is true that the part played by man in
battle was never preponderating or decisive.
Already in the days of Homer, the divinities of Olympus mingled with mortals
on the plains of Troy and, wrapped in
their silvery cloud, which rendered them

invisible without hampering their action, protected, dominated or struck terror into the warriors. But these divinities had a limited power and a limited mystery. Their intervention, although superhuman, still reflected the form and psychology of man. Their secrets revolved in the narrow orbit of our own secrets. The heaven from which they issued was the heaven of our conception; their passions, their sorrows, their thoughts were but little juster, loftier or purer than our own. Then, as man developed, as illusion fell from him, as his consciousness increased and the world stood more plainly revealed, the gods that went with him became greater, although more distant; mightier, though more obscure. With his increase of knowledge and comprehension, the unknown flooded his domain; and, as he organized and extended his armies, perfected his weapons and, with his growing science, mastered natural forces, so do we find the fortune of battle ignoring the captain and heeding only the group of undecipherable laws which we term chance, or hazard, or destiny. Consider, for instance, the

The Gods of War

admirable picture, so palpably true to life, which Tolstoi draws of the battle of Borodino or the Moskowa, a type of one of the great battles of the Empire. The two chiefs, Kutusoff and Napoleon, are so far away from the scene that they perceive only the most insignificant details; they know hardly aught of what is happening. Kutusoff, like the good Sclavonic fatalist that he is, is aware of "the force of circumstance." Sprawling outside a hovel, on a bench over which a carpet has been spread, the unwieldy, one-eyed Russian drowsily awaits the result, giving no orders, content to say "Yes" or "No" to the suggestions that reach him. Napoleon, on the other hand, believes himself able to govern events of which he is not even the witness. He has dictated the arrangements of the battle on the night before; and, from the very first onslaught, owing to that same "force of circumstance" to which Kutusoff pins his faith, not one of these arrangements has been or could have been carried into effect. But he clings none the less to the imaginary plan which reality has shattered; he believes

that he is issuing orders, whereas, in truth,
he is merely following—and that too late—
the mandates of chance that everywhere
arrive in advance of his haggard, hysterical
messengers. And the battle pursues the
course that nature has traced for it, like the
river that flows on its way without heeding
the cries of the men on its banks.

4

And yet Napoleon, of all the generals of
our later wars, remains the only one who
preserved the semblance of human direction. The external forces that seconded and
already dominated the efforts of his troops
were still in their cradle. But, in our day,
what could he do? Would he be able to
recapture one hundredth part of the influence which he was able to exercise on the
fate of battles? For, to-day, the children of
mystery have emerged from childhood; the
gods are other that press on our ranks, break
our lines and scatter our squadrons, sink our
ships and wreck our fortresses. These gods
have no longer a human shape; they issue

from primitive chaos, far beyond the home of
their predecessors; and all their laws, their
power, their intentions must be sought
outside the circle of our own life, on the
other face of our intelligent sphere, in a
world that is closely sealed, the world most
hostile of all to the destinies of our species:
the raw, formless world of inert matter. And
it is to this blind and frightful unknown,
which has nothing in common with us, which
obeys impulses and commands as incomprehensible as those which govern the most
fabulously distant stars; it is to this impenetrable, irresistible energy that we confide
the exclusive attribute of what is highest
in the form of life which we are alone to
represent in this world; it is to these unde-
finable monsters that we entrust the almost
divine mission of establishing the right and
separating the just from the unjust. . . .

5

What are the powers to which we have
thus abandoned our specific privileges? I
think at times of a man whose eyes should

be able to discern what is floating around us, able to distinguish all the population of this ether which our glances assure us to be transparent and empty, even as the blind, did not other senses undeceive them, might hold the darkness to be empty that fastens upon their brow. Suppose such a man to pierce the quicksilver of this crystal sphere which we inhabit and which to us reflects only our own face, our own gestures and our own thoughts. Imagine that, one day, passing beyond the appearances that imprison us, we were to attain at last the essential realities and that the invisible which, on every side, confines us, fells us and lifts us, ordains our retreat, our pause and our advance were suddenly to strip the covering from the immense, the awful, the inconceivable images that, in some hollow of space, must inevitably be borne by the phenomena and laws of nature whereof we are the frail playthings.—Nor should this be looked on merely as a poet's dream; it is now that we dream, when we tell ourselves that these laws have neither face nor form, when we forget so readily

their omnipotent and indefatigable presence;
we are now dreaming the puny dream of
human illusion, whereas then we should
enter the eternal truth of the life without
limit in which our own life is bathed.—The
spectacle would be appalling: it would be a
revelation that would terrify all human
energy and paralyze it at the roots of its
nothingness. Consider, for instance, among
the many illusory triumphs of our blindness,
two fleets that prepare for battle. A few
thousand men, as imperceptible, as helpless,
in their relation to the forces brought into
play, as a handful of ants in a virgin forest;
a few thousand men flatter themselves that
they have enslaved and turned to their purpose, to serve an idea entirely foreign to the
universe, the most immeasurable and the
most dangerous of its laws. Try to provide
each of these laws with an aspect, a physiog-
nomy proportionate and appropriate to its
power and its functions! And, if you
fear to let your mind dwell on what is
impossible and unimaginable, leave out of
count the profoundest, the most august of
these laws, among others that of gravitation,

which the ships obey as well as the sea that
bears them and the earth that bears the sea
and the planets that support the earth. You
would have to seek so far, in such solitudes,
in such infinities, beyond such stars, for the
elements that compose it that the wildest
dream would pause in helplessness, nor the
whole universe suffice to lend a mask.

6

Let us, therefore, take those laws alone
which are more limited, if there be any
that have limits; those which are nearer
to us, if there be any that are near. Let
us take only the laws which these ships
imagine to be submissively confined in their
flanks : the laws which we regard as especially
docile and the daughters of our achievement.
What monstrous form, what gigantic shadow
shall we attribute, to take one instance alone,
to the power of explosives, those recent and
supreme gods, which have just dethroned,
in the temples of war, all the gods of the
past ? With what family of terrors, what unforeseen group of mysteries shall we connect

The Gods of War

them? Melinite, dynamite, panclastite, cordite and roburite, lyddite and ballistite, O ye indescribable spectres, by whose side the old black powder that struck terror into our fathers and even the mighty thunderbolt, once held the most awful symbol of divine anger, become mere gossipy, good-natured old women, a little ready to strike, perhaps, but almost inoffensive, almost maternal: of your countless secrets not even the most superficial has been laid bare; and the chemist who composes your slumber, even as the engineer or artilleryman who awakens you, is in total ignorance of your nature, your origin, your soul, the springs of your incredible bound and the eternal laws which you so suddenly obey! Are you the result of things imprisoned since the beginning of time; are you the gleaming transfiguration of death, the awful gladness of the palpitating void; are you eruption of hatred or excess of joy? Are you a new form of life and so ardent that you consume in a second the patience of twenty centuries? Are you a flame from the enigma of the worlds that has found a

Life and Flowers

fissure in the walls of silence that enclose it?
Are you an audacious loan from the reserve
of energy that supports our earth in space?
Do you, for that unequalled bound of yours
towards a new destiny, gather up, in the
twinkling of an eye, all that has been stored,
all that has been gathered and prepared in
the secret of rocks and seas and mountains?
Are you soul or matter or a third state still
unknown to life? Whence do you derive
your destructive passion, where do you rest
the lever that splits a continent, whence does
the impetus depart that exceeds the zone of
the stars whereon the earth, your mother,
exercises her will?

To all these questions the man of science
who creates you will reply gravely that your
force "is due to the sudden production of a
great volume of gas in a space too confined
to contain it beneath the atmospheric pressure." All is now explained, all is clear.
We attain at once the very depths of truth;
and here, as in all things, know exactly how
matters stand. . .

OUR SOCIAL DUTY

OUR SOCIAL DUTY

LET us start fairly with the great truth:
for those who possess there is only one certain duty, which is to strip themselves of
what they have, so as to bring themselves
into the condition of the mass that possesses
nothing. It is understood, in every clear-
thinking conscience, that no more imperative
duty exists ; but, at the same time, it is admitted that this duty, for lack of courage,
is impossible of accomplishment. For the
rest, in the heroic history of the duties,
even at the most ardent periods, even at
the beginning of Christianity and in the
majority of the religious orders that made a
special cult of poverty, this is perhaps the
only duty that has never been completely
fulfilled. It behoves us, therefore, when

considering our subsidiary duties, to remember that the essential one has been knowingly evaded. Let this truth govern us. Let us not forget that we are speaking in its shadow and that our boldest, our utmost steps will never lead us to the point at which we ought to have been from the first.

2

Since it appears that we have here to do with an absolute impossibility before which it were idle to make any further display of astonishment, let us accept human nature as we find it. Let us, therefore, seek on other roads than the one direct road—seeing that we have not the strength to travel by it— that which, in the absence of this strength, is able to nourish our conscience. There are thus, not to speak of the great question, two or three others which well-disposed hearts are constantly setting to themselves. What are we to do in the actual state of our society? Must we side, *a priori*, systematically, with those who are disorganizing it, or join the camp of those who are struggling

to maintain its economy? Is it wiser not
to bind one's choice, to defend by turns that
which seems reasonable and opportune in
either party? It is certain that a sincere
conscience can find, here or there, the where-
withal to satisfy its activity or to lull its
reproaches. That is why, in the presence of
this choice which to-day becomes incumbent
upon every upright intelligence, it is not
unprofitable to weigh the *pro* and the *contra*
more simply than after our usual fashion
and rather in the manner of the unbiased
denizen of some neighbouring planet.

3

Let us not resume all the traditional
objections, but only those which can be
seriously defended. We are first confronted
with the oldest of them, which maintains
that inequality is inevitable, being in accordance with the laws of nature. This is true;
but the human race appears not improbably
created to raise itself above certain of the
laws of nature. Its very existence would be
imperilled if it abandoned its intention to

surmount a number of these laws. It is in accordance with its particular nature to obey other laws than those of its animal nature and the rest. Moreover, this objection has long been classed among those whose principle is untenable and would lead to the massacre of the weak, the sick, the old and so forth.

We are next told that it is right, in order to hasten the triumph of justice, that the best among us should not prematurely strip themselves of their arms, the most efficacious of which are exactly wealth and leisure. Here the necessity of the great sacrifice is fairly well recognized and only the question of its opportuneness remains. We agree, provided that it be well understood that this wealth and leisure serve solely to hasten the steps of justice.

Another conservative argument worthy of attention declares that, man's first duty being to avoid violence and bloodshed, it is indispensable that the social evolution should not be too rapid, that it should ripen slowly, that it is important to temper it while the mass is being enlightened and borne gradually and

without serious upheavals towards a liberty
and a fulness of possessions which, at this
moment, would unchain only its worst instincts. This again is true; nevertheless, it
would be interesting to calculate, since we
can reach the best only through the bad,
whether the evils of a sudden, radical and
bloody revolution outweigh those which are
perpetuated in the slower evolution. It were
well to ask ourselves whether there be not an
advantage in acting with all speed; whether,
when all is told, the suffering of those who
now wait for justice be not more serious than
that which the privileged class of to-day
would have to undergo for the space of some
weeks or months. We are too ready to forget
that the headsmen of misery are less noisy,
less theatrical, but infinitely more numerous,
cruel and active than those of the most
terrible revolutions.

4

We come at length to the last and perhaps
the most disturbing argument: humanity, we
are told, has for more than a century been
passing through the most fruitful and

victorious, probably the climacteric years of
its destiny. It seems, if we consider its past,
to be in the decisive phase of its evolution.
One would think, from certain indications,
that it is nigh upon attaining its apogee.
It is traversing a period of inspiration wherewith none other is historically to be compared.
A trifle, a last effort, a flash of light which
shall connect or emphasize the discoveries,
the intuitions scattered or held in suspense
alone separates it, perhaps, from the great
mysteries. It has lately touched upon
problems whose solution, at the cost of the
hereditary enemy, that is of the great unknown phenomenon of the universe, would
probably render useless all the sacrifices
which justice demands of men. Is it not
dangerous to stop this flight, to disturb this
precious, precarious and supreme minute?
Admitting even that what is gained can no
longer be lost, as in the earlier upheavals,
it is nevertheless to be feared lest the vast
disorganization required by equity should
put an abrupt end to this happy period;
and it is not sure but that its reappearance
might be long delayed, the laws which

preside over the inspiration of the genius
of the race being as capricious, as unstable as
those which preside over the inspiration of
the genius of the individual.

5

This is, as I have said, perhaps the most
disquieting argument. But there is no
doubt that it attaches too great an importance to a somewhat uncertain danger.
Moreover, prodigious compensations would
attend this brief interruption of the victory
of humanity. Can we foresee what will
happen when the human race as a whole
will be taking part in the intellectual labour
which is the labour proper to our species?
To-day, hardly one brain in ten thousand
exists in conditions entirely favourable to
its activity. There is, at this moment, a
monstrous waste of spiritual force. Idle-
ness at the top depresses as many mental
energies as excess of manual labour anni-
hilates below. It is incontestable that, when
it shall be given to all men to apply themselves to the task at present reserved for a

few favourites of chance, humanity will
increase a thousandfold its prospects of
attaining the great mysterious aim.

Here, I think, we have the best of the
pro and the *contra*, the most reasonable
reasons that can be invoked by those who
are in no hurry to end the matter. In the
midst of these reasons stands the huge
monolith of injustice. There is no need
to let it defend itself. It oppresses con-
sciences, limits intelligences. Wherefore
there can be no question of not destroying
it; all that is asked of those who would
overthrow it is a few years of patience, so
that, when its surroundings have been
cleared, its fall may entail fewer disasters.
Are we to grant these years? And which
among these arguments in favour of haste
or of waiting would be the object of the
most straightforward choice?

6

Do the pleas for a few years of respite
appear to you sufficient? They are precarious enough; but, even so, it would not

be fair to condemn them without considering the problem from a higher standpoint
than that of pure reason. This 'point must
always be sought as soon as we have to do
with questions that go beyond human experience. It might easily be maintained,
for instance, that the choice would not be
the same for all. The race, which probably
has an infinite consciousness of its destinies
which no individual can grasp, would have
very wisely apportioned among men the
parts that suit them in the lofty drama of
its evolution. For reasons which we do not
always understand, it is doubtless necessary
that the race should progress slowly: that
is why the enormous mass of its body attaches it to the past and the present; and
very upright intelligences may be comprised
within this mass, even as it is possible for
greatly inferior minds to escape from it.
Whether there be satisfaction or unselfish
discontent on the side of the darkness or of
the light matters little; it is often a question of predestination and the distribu-
tion of characters rather than of enquiry.
However this may be, for us, whose reason

already judges the weakness of the arguments of the past, it would be a fresh motive for impatience. Let us admit, in addition, its very plausible force. The fact, therefore, that to-day does not satisfy us is enough to make it our duty, our organic duty, so to speak, to destroy all that supports it, in order to make ready for the arrival of to-morrow. Even if we were to perceive very clearly the dangers and drawbacks of too prompt an evolution, it is requisite, in order that we should loyally fulfil the function assigned to us by the genius of the race, that we should take no notice of any patience, any circumspection. In the social atmosphere, we represent the oxygen : if we behave in it like the inert azote, we betray the mission which nature has entrusted to us ; and this, in the scale of the crimes that remain to us, is the gravest and most unpardonable of treasons. It is not ours to preoccupy our minds with the often grievous consequences of our haste : this is not written in our part and to take account of it would be to add to that part discordant words which are not in the authentic text

Our Social Duty

dictated by nature. Humanity has appointed us to gather that which stands on the horizon. It has given us instructions which it does not behove us to discuss. It distributes its forces as it thinks right. At every cross-way on the road that leads to the future, it has placed, against each of us, ten thousand men to guard the past; let us therefore have no fear lest the fairest towers of former days be insufficiently defended. We are only too naturally inclined to temporize, to shed tears over inevitable ruins: this is the greatest of our trespasses. The least that the most timid among us can do —and already they are very near committing treachery—is not to add to the immense dead - weight which nature drags along. But let the others follow blindly the inmost impulse of the power that urges them on. Even if their reason were to approve none of the extreme measures in which they take part, let them act and hope beyond their reason; for in all things, because of the call of the earth, we must aim higher than the object which we aspire to attain.

Life and Flowers

7

Let us not fear lest we be drawn too far; and let no reflection, however just, break or temper our ardour. Our future excesses are essential to the equilibrium of life. There are men enough about us whose exclusive duty, whose most precise mission it is to extinguish the fires which we kindle. Let us go always to the most extreme limits of our thoughts, our hopes and our justice. Let us not persuade ourselves that these efforts are incumbent only upon the best of us: this is not true and the humblest among us that foresee the coming of a dawn which they do not understand must await it at the very summit of themselves. Their presence on these intermediary tops will fill with living substance the dangerous intervals between the first heights and the last and will maintain the indispensable communications between the vanguard and the mass.

Let us think sometimes of the great invisible ship that carries our human destinies upon eternity. Like the vessels of our confined oceans, she has her sails and her ballast.

Our Social Duty

The fear that she may pitch or roll on leaving the roadstead is no reason for increasing the weight of the ballast by stowing the fair, white sails in the depths of the hold. They were not woven to moulder side by side with cobble-stones in the dark. Ballast exists everywhere: all the pebbles of the harbour, all the sand on the beach will serve for it. But sails are rare and precious things : their place is not in the murk of the well, but amid the light of the tall masts, where they will collect the winds of space.

8

Let us not say to ourselves that the best truth always lies in moderation, in the decent average. This would perhaps be so if the majority of men did not think, did not hope upon a much lower plane than is needful. That is why it behoves the others to think and hope upon a higher plane than seems reasonable. The average, the decent moderation of to-day will be the least human of things to-morrow. At the time of the Spanish Inquisition, the opinion of good

sense and of the just medium was certainly
that people ought not to burn too large a
number of heretics; extreme and unreasonable opinion obviously demanded that they
should burn none at all. It is the same
to-day with the question of marriage, of
love, of religion, of criminal justice and so
on. Has not mankind yet lived long enough
to realize that it is always the extreme idea,
that is the highest idea, the idea at the
summit of thought, that is right? At the
present moment, the most reasonable opinion
on the subject of our social question invites
us to do all that we can gradually to diminish inevitable inequalities and distribute
happiness more equitably. Extreme opinion
demands instantly integral division, the suppression of property, obligatory labour and
the rest. We do not yet know how these
demands will be realized; but it is already
quite certain that very simple circumstances
will one day make them appear as natural as
the suppression of the right of primogeni-
ture or of the privileges of the nobility. It
is important, in these questions of the dura-
tion of a species and not of a people or an

Our Social Duty

individual, that we should not limit ourselves to the experience of history.   Anything
that it confirms or denies moves in an insignificant circle.   The truth, in this case, lies
much less in our reason, which is always
turned towards the past, than in our imagination, which sees farther than the future.

9

Let our reason, then, strive to soar above
experience.   This is easy for young people;
but it is salutary that ripe age and old age
should learn to raise themselves to the
luminous ignorance of youth.   We should
guard beforehand, as the years pass, against
the dangers which our confidence in the race
must run because of the great number of
malignant men whom we have encountered
in it.   Let us continue, in spite of all, to
act, to love and to hope as though we
had to do with an ideal humanity.   This
ideal is only a vaster reality than that
which we behold.   The failings of individuals no more impair the general purity
and innocence than the waves on the

surface, according to the aeronauts, when
seen from a certain height, trouble the
profound limpidity of the sea.

10

Let us listen only to the experience that
urges us on; it is always higher than that
which throws or keeps us back. Let us reject
all the counsels of the past that do not turn
us towards the future. This is what was
admirably understood, perhaps for the first
time in history, by certain men of the French
Revolution; and that is why this revolution
is the one that did the greatest and the most
lasting things. Here this experience teaches
us that, contrary to all that occurs in the
affairs of daily life, it is above all important
to destroy. In every social progress, the
great and the only difficult work is the destruction of the past. We need not be
anxious about what we shall place in the
stead of the ruins. The force of things and
of life will undertake the rebuilding. It is
but too eager to reconstruct; and we should
not be doing well to aid it in its precipitate

task. Let us, therefore, not hesitate to employ our destructive powers even to excess : nine-tenths of the violence of our blows is lost amid the inertness of the mass, even as the stroke of the heaviest hammer is dispersed in a large stone and becomes, so to speak, imperceptible to a child that holds the stone in its hand.

II

And let us not fear lest we should go too fast. If, at certain hours, we seem to be rushing at a headlong and dangerous pace, this is to counterbalance unjustifiable delays and to make up for time lost during centuries of inactivity. The evolution of our world continues during these periods of inertia ; and it is probably necessary that humanity should have reached a certain determined point of its ascent at the moment of a certain sidereal phenomenon, of a certain obscure crisis of the planet, or even of the birth of a certain man. It is the instinct of the race that decides

these matters, it is its destiny that speaks;
and, if this instinct or this destiny be
wrong, it is not for us to interfere; for
there is nothing above it or above ourselves
to correct its error.

OUR ANXIOUS MORALITY

OUR ANXIOUS MORALITY

I

WE have arrived at a stage of human evolution that must be almost unprecedented in history. A large portion of mankind—and just that portion which corresponds with the part that has hitherto created the events of which we know with some certainty—is gradually forsaking the religion in which it has lived for nearly twenty centuries.

For a religion to become extinct is no new thing. It must have happened more than once in the night of time; and the annalists of the end of the Roman Empire make us assist at the death of paganism. But, until now, men passed from a crumbling temple into one that was building; they left one religion to enter another; whereas we are abandoning ours to go nowhither.

Life and Flowers

That is the new phenomenon, with the unknown consequences, wherein we live.

2

It is not necessary to recall the fact that religions have always, through their morality and their promises extending beyond the tomb, exercised an enormous influence upon men's happiness, although we have seen some—and very important ones, such as paganism — which provided neither those promises nor any morality properly so-called. We will not speak of the promises of our own religion, for they are the first to perish with the faith, whereas we are still living in the monuments erected by the morality born of that departing faith. But we feel that, in spite of the supports of habit, these monuments are yawning over our heads and that already, in many places, we are shelterless under an unforeseen heaven that has ceased to give its orders. Thus we are assisting at the more or less unconscious and feverish elaboration of a morality that is premature, because we feel it to be indispensable, made up of remnants gathered from the past, of conclusions borrowed from ordinary good sense, of a few laws half perceived by science and, lastly, of certain extreme intuitions of our bewildered intelligence, which returns, by a circuitous road through a new mystery, to old-time virtues which good sense alone is not sufficient to sustain. It may be interesting to try to seize the first reflexes of that elaboration. The hour seems to be striking at which many will ask themselves whether, by continuing to practise a lofty and noble morality in an environment that obeys other laws, they be not disarming themselves too artlessly and playing the ungrateful part of dupes. They wish to know if the motives that still attach them to the older virtues are not merely senti-mental, traditional and illusionary; and they seek somewhat vainly within themselves for the supports which reason may yet lend them.

3

Placing on one side the artificial heaven in which those who remain faithful to the

religious certainties take shelter, we find that
the upper currents of civilized humanity
waver, seemingly, between two contrary doctrines. For that matter, these two parallel,
but inverse doctrines have through all time,
like hostile streams, crossed the fields of
human morality. But their bed was never
so clearly, so rigidly dug out as now. That
which in other days was no more than altruism and egoism instinctive and vague,
with waves that often mingled, has of late
become altruism and egoism absolute and
systematic. At their sources, which are not
renewed, but shifted, stand two men of
genius : Tolstoi and Nietzsche. But, as I
have said, it is only seemingly that these
two doctrines divide the world of ethics.
The real drama of the modern conscience
is not enacted at either of these too extreme
points. Lost in space, they mark little
more than two illusive goals, which nobody
dreams of attaining. One of these doctrines
flows violently back towards a past that
never existed in the shape in which that
doctrine pictures it ; the other ripples cruelly
towards a future which there is nothing

to foretell. Between these two dreams, which envelop and go beyond it on every side, passes the reality of which they have failed to take account. In this reality, whereof each of us carries the image within himself, it behoves us to study the formation of the morality on which our latter-day life rests. Need I add that, when employing the term "morality," I do not mean to speak of the practices of daily existence, which spring from custom and fashion, but of the great laws that determine the inner man?

4

Our morality is formed in our conscious or unconscious reason, which, from this point of view, may be divided into three regions. Right at the bottom lies the heaviest, the densest and the most general, which we will call "common sense." A little higher, already striving towards ideas of immaterial usefulness and enjoyment, is what might be called "good sense." Lastly, at the top, admitting, but controlling as severely as possible the claims of the

imagination, of the feelings and of all that
connects our conscious life with the unconscious and with the unknown forces within
and without, lies the indeterminate part of
that same total reason, to which we will give
the name of " mystic reason."

5

It is not necessary to set forth at length
the morality of "common sense," of that
good common sense which exists in all of us,
in the best and the worst of us alike, and
which springs up spontaneously on the ruins
of the religious idea. It is the morality of
each man for himself, of practical, solid
egoism, of every material instinct and enjoyment. He who starts from common
sense considers that he possesses but one
certainty: his own life. In that life, going
to the bottom of things, are but two real
evils: sickness and poverty; and but two
genuine and irreducible boons: health and
riches. All other realities, happy or unhappy, flow from these. The rest—joys and
sorrows born of the feelings and the passions

Our Anxious Morality

—is imaginary, because it depends upon the idea which we form of it. Our right to enjoyment is limited only by the similar right of those who live at the same time as ourselves; and we have to respect certain laws established in the very interest of our peaceful enjoyment. With the reservation of these laws, we admit no constraint; and our conscience, so far from trammelling the movements of our selfishness, must, on the contrary, approve of their triumphs, seeing that these triumphs are what is most in accordance with the instinctive and logical duties of life.

There we have the first stratum, the first state of all natural morality. It is a state beyond which many men, after the complete death of the religious ideas, will never go.

6

As for "good sense," which is a little less material, a little less animal, it looks at things from a slightly higher standpoint and, consequently, sees a little farther. It soon perceives that niggardly common sense

leads an obscure, confined and wretched life
in its shell. It observes that man is no more
able than the bee to remain solitary and that
the life which he shares with his fellows,
in order to expand freely and completely,
cannot be reduced to an unjust and pitiless
struggle or to a mere exchange of services
grudgingly rewarded. In its relations towards others, it still makes selfishness its
starting-point ; but this selfishness is no
longer purely material. It still considers
utility, but already admits its spiritual or
sentimental side. It knows joys and sorrows, affections and antipathies, the objects
of which may exist in the imagination.
Thus understood and capable of rising to
a certain height above the conclusions of
material logic—without losing sight of its
interest—it appears beyond the reach of
every objection. It flatters itself that it is
in solid occupation of all reason's summits.
It even makes a few concessions to that which
does not perceptibly fall within the latter's
domain, I mean to the passions, the feelings
and all the unexplained things that surround them. It must needs make these

concessions, for, if not, the gloomy caves in which it would shut itself would be no more habitable than those in which dull common sense leads its stupefied existence. But these very concessions call attention to the unlawfulness of its claims to busy itself with morality once that the latter has gone beyond the ordinary practices of daily life.

7

Indeed, what can there be in common between good sense and the stoical idea of duty, for instance? They inhabit two different and almost uncommunicating regions. Good sense, when it claims alone to promulgate the laws that form the inner man, ought to meet with the same resistance and the same obstacles as those against which it strikes in one of the few regions which it has not yet reduced to slavery: the region of æsthetics. Here it is very happily consulted on all that concerns the starting-point and certain great lines, but most imperiously ordered to hold its peace so soon as the achievement and the supreme and mysterious beauty of the work come into question. But, whereas in æsthetics it resigns itself easily enough to silence, in morality it wishes to lord over all things. It were well, therefore, to put it back once for all into its lawful place in the generality of the faculties that make up our human person.

8

One of the features of our time is the ever-increasing and almost exclusive confidence which we place in those parts of our intelligence which we have just described as common sense and good sense. It was not always thus. Formerly, man based upon good sense only a somewhat restricted and the vulgarest portion of his life. The rest had its foundations in other regions of our mind, notably in the imagination. The religions, for instance, and with them the brightest part of the morality of which they are the chief sources, always rose up at a great distance from the tiny limits of good sense. This was excessive; but the question is whether the present contrary excess

is not as blind. The enormous strides made
in the practice of our life by certain mechanical and scientific laws make us allow to
good sense a preponderance to which it
remains to be proved that this same good
sense is entitled. The apparently incontestable, yet perhaps illusory logic of certain
phenomena with which we believe ourselves
acquainted makes us forget the possible
illogicality of millions of other phenomena
which we do not yet know. Nothing
assures us that the universe obeys the laws
of human logic. It would even be surprising if this were so; for the laws of
our good sense are the fruit of an experience which is insignificant when we compare
it with what we do not know. " There is
no effect without a cause," says our good
sense, to take the tritest instance. Yes,
in the little circle of our material life,
that is undeniable and all-sufficing. But, so
soon as we emerge from this infinitesimal
circle, the saying no longer answers to anything, seeing that the notions of cause and
effect are alike unknowable in a world where
all is unknown. Now our life, from the

moment when it raises itself a little, is constantly issuing from the small material and experimental circle and, consequently, from the domain of good sense. Even in the visible world which serves it for a model in our mind, we do not observe that it reigns undivided. Around us, in her most constant and most familiar manifestations, nature very rarely acts according to good sense. What could be more senseless than her waste of existences? What more unreasonable than those billions of germs blindly squandered to achieve the chance birth of a single being? What more illogical than the untold and useless complication of her means (as, for instance, in the life of certain parasites and the impregnation of flowers by insects) to attain the simplest ends? What madder than those thousands of worlds which perish in space without accomplishing a single work? All this goes beyond our good sense and shows it that it is not in agreement with general life and that it is almost isolated in the universe. Needs must it argue against itself and recognize that we shall not give

it in our life, which is not isolated, the
preponderant place to which it aspires. This
is not to say that we will abandon it where
it is of use to us; but it is well to know
that good sense cannot suffice for everything, being itself almost nothing. Even as
there exists without ourselves a world that
goes beyond it, so there exists within ourselves
another that exceeds it. It is in its place
and performs a humble and blessed work in
its little village; but it must not aim at
becoming the master of the great cities and
the sovereign of the mountains and the seas.
Now the great cities, the seas and the mountains occupy infinitely more space within us
than the little village of our practical existence, which is the necessary agreement upon
a small number of inferior, sometimes doubtful, but indispensable truths and nothing
more. It is a bond rather than a support.
We must remember that nearly all our progress has been made in despite of the sarcasms
and curses with which good sense has received
the unreasonable, but fertile hypotheses of the
imagination. Amid the moving and eternal
waves of a boundless universe, let us not,

therefore, hold fast to our good sense as
though to the one rock of salvation. Bound
to that rock, immovable through every age
and every civilization, we should do nothing
of that which we ought to do, become
nothing of that which we may perhaps
become.

9

Until the present time, this question of a
morality limited by good sense possessed no
great importance. It did not stay the development of certain aspirations, of certain
forces that have always been considered the
finest and noblest to be found in man. The
religions completed the interrupted work.
To-day, feeling the danger of its limitations,
the morality of good sense, which would like
to become the general morality, seeks to
extend itself as far as possible in the direction of justice and generosity, to find, in a
superior interest, reasons for being disinterested, in order to fill up a portion of the
abyss that separates it from those indestructible forces and aspirations. But there are
points which it is unable to exceed without

denying itself, without destroying itself at
its very source. After these points, which
are just those at which the great useless
virtues begin, what guide remains to us?

10

We shall see presently if it be possible to
answer this question. But, even admitting
that there is not, that there never can be a
guide beyond the plains of the morality of
good sense, this is no reason why we should
be anxious touching the moral future of
humanity. Man is so essentially, so necessarily a moral being that, when he denies
the existence of all morality, that very denial
already becomes the foundation of a new
morality. Mankind, at a pinch, can do
without a guide. It proceeds a little more
slowly, but almost as surely through the
darkness which no one lights. It carries
within itself the light whose flame is blown
to and fro, but incessantly revived by the
storms. It is, so to speak, independent of
the ideas which imagine that they lead it.
Moreover, it is interesting and easy to

establish, that these periodical ideas have
always had but little influence on the mass
of good and evil that is done in the world.
The only thing that has a real influence is the
spiritual wave which carries us, which has its
ebbs and flows, but which seems slowly to
overtake and conquer we know not what in
space. More important than the idea is
the time that lapses around it, is the development of a civilization which is only
the level of the general intelligence at a
given moment in history. If a religion
were revealed to us to-morrow, proving,
scientifically and with absolute certainty,
that every act of goodness, of self-sacrifice,
of heroism, of inward nobility would bring
us, immediately after our death, an indubitable and unimaginable reward, I doubt
whether the proportion of good and evil,
of virtues and vices amid which we live
would undergo an appreciable change.
Would you have a convincing example?
In the middle ages there were moments
when faith was absolute and obtruded itself
with a certainty that corresponds exactly with
our scientific certainties. The rewards pro-

mised for well-doing, the punishments threat-
ening evil were, in the thoughts of the men of
that time, as tangible, so to speak, as would
be those of the revelation of which I spoke
above. Nevertheless, we do not see that
the average of goodness was raised. A few
saints sacrificed themselves for their brothers,
carried certain virtues, selected from among
the more contestable, to the pitch of heroism ;
but the bulk of men continued to deceive
one another, to lie, to fornicate, to steal, to
be guilty of envy, to commit murder. The
mean of the vices was no lower than that of
to-day. On the contrary, life was incom-
parably harsher, more cruel and more unjust,
because the low-water mark of the general
intelligence was less high.

I I

Let us return to our positivist, utilitarian,
materialist or rational morality, which we
have called the morality of common sense
and good sense. It is certain that, beside the
latter, there has always been, there still is
another which embraces all that extends from

the virtues of good sense, which are necessary
to our material and spiritual happiness, to
the infinity of heroism, of self-sacrifice, of
goodness, of love, of inward probity and
dignity. It is certain that the morality of
good sense, although it may go pretty far in
some directions, such as that of altruism, for
instance, will always be a little wanting in
nobility, in disinterestedness and, above all,
in I know not what faculties that are capable
of bringing it into direct relations with the
uncontested mystery of life.

If it be probable, as we have hinted, that
our good sense answers only to an infinitesimal portion of the phenomena, the truths
and the laws of nature, if it isolate us somewhat piteously in this world, we have within
us other faculties which are marvellously
adapted to the unknown parts of the universe and which seem to have been given to
us expressly to prepare us, if not to understand them, at least to admit them and to
undergo their great presentiments. These
are imagination and the mystic summit of
our reason. Do and say what we may, we
have never been, we are not yet a sort of

Our Anxious Morality

purely logical animal. There is in us, above the reasoning portion of our reason, a whole region which answers to something different, which is preparing for the surprises of the future, which is awaiting the events of the unknown. This part of our intelligence, which I will call imagination or " mystic reason," went before us in times when, so to speak, we knew nothing of the laws of nature, outran our imperfect attainments and made us live, morally, socially and sentimentally, on a level very much superior to that of those attainments. At the present time, when we have made the latter take a few steps forward in the darkness and when, in the hundred years that have just elapsed, we have unravelled more chaos than in a thousand previous centuries, at the present time, when our material life seems on the point of becoming fixed and assured, is this a reason why these two faculties should cease to go ahead of us or why they should retrocede towards good sense? Are there not, on the contrary, very serious reasons for urging them forwards, so as to restore the normal distances and their traditional lead? Is it right that we should lose confidence in them? Is it possible to say that they have hindered any form of human progress? Perhaps they have deceived us more than once; but their fruitful errors, by forcing us to march onwards, have revealed to us, in the straying, more truths than our over-timid good sense would ever have lighted upon by marking time. The most welcome discoveries, in biology, in chemistry, in medicine, in physics, almost all had their starting-point in an hypothesis supplied by imagination or mystic reason, an hypothesis which the experiments of good sense have confirmed, but which, given as it is to narrow methods, it would never have foreseen.

12

In the exact sciences, in which it seems as if they ought to be first dethroned, imagination and mystic reason (that is to say that part of our reason which extends above good sense, draws no conclusions and plays an enormous and lawful part in the hesitations and possibilities of the unknown), our imagination, I was saying, and our mystic reason again occupy a place of honour. In æsthetics, they reign almost undivided. Why should silence be laid upon them in our morality, which fills an intermediary space between the exact sciences and æsthetics? There is no concealing the fact : if they cease to come to the assistance of good sense, if they give up prolonging its work, the whole summit of our morality falls in abruptly. Starting from a certain line which is exceeded by the heroes, by the great wise men and even by the majority of mere good men, all the height of our morality is the fruit of our imagination and belongs to mystic reason. The ideal man, as formed by the most enlightened and the most extensive good sense, does not yet correspond, does not even correspond in the slightest degree with the ideal man of our imagination. The latter is infinitely higher, more generous, nobler, more disinterested, more capable of love, of self-abnegation, of devotion and of essential sacrifices. It is a question of knowing which of the two is right or wrong, which has the right of surviving.

Or, rather, it is a question of knowing whether some new fact permit us to make this demand and to bring into question the high traditions of human morality.

13

Where shall we find this new fact? Among all the revelations which science has lately given us, is there a single one that authorizes us to take anything from the ideal set before us by Marcus Aurelius, for instance? Does the least sign, the least indication, the least presentiment arouse a suspicion that the primitive ideas which hitherto have guided the just man will have to change their direction and that the road of human good-will is a false road? What discovery tells us that it is time to destroy in our conscience all that goes beyond strict justice, that is to say those unnamed virtues which, beyond those necessary to social life, appear to be weaknesses and yet turn the simple decent man into the real and profound good man?

Those virtues, we shall be told, and a host

of others that have always formed the perfume of great souls, those virtues would doubtless be in their places in a world where the struggle for life was no longer so necessary as it is now on a planet on which the evolution of species is not yet finished. Meanwhile, most of them disarm those who practise them as against those who do not practise them. They trammel the development of those who ought to be the best to the advantage of the less good. They oppose an excellent, but human and particular ideal to the general ideal of life; and this more restricted ideal is necessarily vanquished beforehand.

The objection is a specious one. First of all, this so-called discovery of the struggle for life, in which men seek the source of a new morality, is at bottom but a discovery of words. It is not enough to give an unaccustomed name to an immemorial law in order to render lawful a radical deviation from the human ideal. The struggle for life has existed since the existence of our planet; and not one of its consequences was modified, not one of its riddles solved on the day when men thought that they had taken cognizance of it by adorning it with an appellation which a whim of the vocabulary will change, perhaps, before fifty years have passed. Next, it behoves us to admit that, if these virtues sometimes disarm us in the face of those who do not know them, they disarm us only in very contemptible combats. Certainly, the over-scrupulous man will be deceived by him who is unscrupulous, the too-loving, over-indulgent, too-devoted man will suffer at the hands of him who is less so; but can this be called a victory of the second over the first? In what does this defeat strike at the inner life of the better man? He will lose some material advantage by it; but he would lose much more by leaving uncultivated all the region that extends beyond the morality of good sense. The man who enriches his sensibility enriches his intelligence; and these are the properly human forces that always end by having the last word.

Our Anxious Morality

14

Moreover, if a few general thoughts succeed in emerging from the chaos of half-discoveries, of half-truths that beguile the mind of modern man, does not one of these thoughts assert that nature has given to every species of living being all the instincts necessary for the accomplishment of its destinies? And has she not, at all times, given us a moral ideal which, in the most primitive savage and the most refined civilized man alike, preserves a proportional and perceptibly equal distance ahead of the conclusions of good sense? Is not the savage, just as, in a higher sphere, the civilized man, as a rule infinitely more generous, more loyal, more true to his word than the interest and experience of his wretched life advise? Is it not thanks to this instinctive ideal that we live in an environment in which, despite the practical preponderance of evil, excused by the harsh necessities of existence, the idea of goodness and justice reigns more and more supreme and in which the public conscience, which is the perceptible and general form of that ideal, becomes more and more powerful and certain of itself?

15

It is fitting that we should come to an understanding, once for all, on the rights of our instincts. We no longer allow the rights of any of our lower instincts to be contested. We know how to justify and to ennoble them by attaching them to some great law of nature. Why should not certain more elevated instincts, quite as incontestable as those which crawl at the bottom of our senses, enjoy the same pre-rogatives? Must they be denied, suspected or treated as illusions because they are not related to the two or three primitive neces-sities of animal life? Once that they exist, is it not probable that they are as indispen-sable as the others to the accomplishment of a destiny concerning which we do not know what is useful or useless to it, seeing that we do not know its objects? And is it not, then, the duty of our good sense, their

innate enemy, to help them, to encourage
them and, finally, to confess to itself that
certain parts of our life are beyond its
sphere ?

It is our duty, above all, to strive to
develop within ourselves the specific characteristics of the class of living being to
which we belong and, by preference, those
which distinguish us the most from all the
other phenomena of the life around us.
Among these characteristics, one of the most
notorious is, perhaps, not so much our
intelligence as our moral aspirations. One
portion of these aspirations emanates from
our intelligence ; but another has always
gone before it, has always appeared independent of it and, finding no visible roots
in it, has sought elsewhere, no matter
where, but especially in the religions the
explanation of a mysterious instinct that
urged it to go farther. To-day, when the
religions are no longer qualified to explain
anything, the fact none the less remains ;
and I do not think that we have the right

to suppress with a stroke of the pen a whole region of our inner existence with the sole object of gratifying the reasoning organs of our judgment. Besides, all things hang together and help one another, even those which seem to contend with one another, in the mystery of man's instincts, faculties and aspirations. Our intelligence derives an immediate profit from the sacrifices which it makes to our imagination when the latter caresses an ideal which the former does not think consonant with the realities of life. Our intelligence has for some years been too prone to believe that it is able to suffice for itself. It needs all our forces, all our feelings, all our passions, all our unconsciousness, all that is with it and all that is against it, in order to spread and flourish in life. But the nutriment which is necessary to it above any other is the great anxieties, the grave sufferings, the noble joys of our heart. These truly are to it what the water from heaven is to the lilies, the dew of the morning to the roses. It is well that it should know how to stoop and pass in silence before certain desires and certain dreams of that heart which it does not always understand, but which contains a light that has more than once led it towards truths which it sought in vain at the extreme points of its thoughts.

17

We are an indivisible spiritual whole; and it is only for the needs of the spoken or written word that we are able, when we study them, to separate the thoughts of our intelligence from the passions and sentiments of our heart. Every man is more or less the victim of this illusory division. He says to himself, in his youth, that he will see into it more clearly when he is older. He imagines that his passions, even the most generous of them, obscure and disturb his thought; and he asks himself, with I know not what hope, how far that thought will go when it reigns alone over his lulled dreams and senses. And old age comes: the intelligence is clear, but has no object remaining. It has nothing left to do, it works in the void. And it is thus that, in the domains where the results of that division are the most visible, we observe that, in general, the work of old age is not equal to that of youth or of mature age, which, nevertheless, has much less experience and knows many fewer things, but which has not yet stifled the mysterious forces foreign to our intelligence.

18

If we be now asked which, when all is said, are the precepts of that lofty morality whereof we have spoken without defining it, we will reply that it presupposes a state of soul or of heart rather than a code of strictly-formulated precepts. What constitutes its essence is the sincere and strong wish to form within ourselves a powerful idea of justice and of love that always rises above that formed by the clearest and most generous portions of our intelligence. One could mention a thousand examples : I will take one only, that which is at the centre of all our anxieties and beside which all the rest has no importance, that which, when we thus speak of lofty and noble morality and perfect virtues, cross-examines us as culprits and asks us bluntly, "And when do you intend to put a stop to the injustice in which you live?"

Yes, all of us who possess more than the others, we who are more or less rich as against those who are quite poor, we all live in the midst of an injustice deeper than that which arises from the abuse of brute strength, because we abuse a strength which is not even real. Our reason deplores this injustice, but explains it, excuses it and declares it to be inevitable. It shows us that it is impossible to apply to it the swift and efficacious remedy which our equity seeks; that any too radical remedy would carry with it evils more cruel and more desperate than those which it pretended to cure; it proves to us, in short, that this injustice is organic, essential and in conformity with all the laws of nature. Our reason is perhaps right; but what is much more deeply, much more surely right is our ideal of justice, which proclaims that our reason is wrong. Even when it is not

acting, it is well, if not for the present,
at least for the future, that this ideal should
have a quick sense of iniquity; and, if it
no longer involves renunciations or heroic
sacrifices, this is not because it is less noble
or less sure than the ideal of the best religions, but because it promises no other
rewards than those of duty accomplished
and because these rewards are just those
which hitherto only a few heroes have
understood and which the great presentiments that hover beyond our intelligence
are seeking to make us understand.

19

In reality, we need so few precepts! . . .
Perhaps three or four, at the utmost five
or six, which a child could give us. We
must, before all, understand them; and " to
understand," as we take it, is hardly, as a
rule, the beginning of the life of an ideal.
If that were enough, all our intelligences
and all our characters would be equal; for
every man of even a very mean intelligence
is apt to understand, at this first stage, all

that is explained to him with sufficient clear-
ness. There are as many manners and as
many stages in the manners of understanding a truth as there are minds that think
that they understand it. If I prove, for
instance, to an intelligent vain man how
childish is his vanity, to an egoist capable
of comprehension how unreasonable and
hateful is his egoism, they will readily
agree, they will even amplify what I have
said. There is, therefore, no doubt that
they have understood; but it is very nearly
certain that they will continue to act as
though not so much as the extremity of
one of the truths which they have just
admitted had grazed their brain. Whereas,
in another man, these truths, covered with
the same words, will one evening suddenly
enter and penetrate, through his thoughts,
to the very bottom of his heart, upsetting
his existence, displacing every axis, every
lever, every joy, every sorrow, every object
of his activity. He has understood the
sense of the word "to understand," for
we cannot flatter ourselves that we have
understood a truth until it is impossible

for us not to shape our lives in accordance
with it.

20

To return to and resume the central
idea of all this, let us recognize that it is
necessary to maintain the equilibrium between what we have called good sense and
the other faculties and sentiments of our
life. Contrary to our former wont, we
are nowadays too much inclined to shatter
this equilibrium in favour of good sense.
Certainly, good sense has the right to control more strictly than ever all that other
forces bring to it, all that goes beyond the
practical conclusions of its reasoning; but
it cannot prevent them from acting until
it has acquired the certainty that they are
deceiving it; and it owes to itself, to the
respect of its own laws the duty of being
more and more circumspect in asserting
that certainty. Now, though it may have
acquired the conviction that those forces
have committed a mistake in ascribing to a
divine and precise will and injunctions the
majority of the phenomena manifested within

Our Anxious Morality

themselves; though it be its duty to redress
the accessory errors that proceed from this
central error, by eliminating, for instance,
from our moral ideal a host of sterile and
dangerous virtues, it could never deny that
these same phenomena subsist, whether they
emanate from a superior instinct, from the
life of the species, infinitely more powerful
within us than the life of the individual, or
from any other unintelligible source. In
any case, it could not treat them as illusions; for, at that rate, we might ask ourselves whether this supreme judge, out-
flanked and contradicted on every side by
the genius of nature and the inconceivable
laws of the universe, be not itself more
illusive than the illusions which it aspires
to destroy.

21

For all that touches upon our moral life,
we still have the choice of our illusions:
good sense itself, that is to say the scientific
spirit, is obliged to admit as much. Wherefore, taking one illusion with another, let us
welcome those from above rather than those

from below. The former, after all, have brought us to the stage at which we are; and, when we look back upon our starting-point, the dreadful cave of prehistoric man, we owe them a certain gratitude. The latter illusions, those of the inferior regions, that is to say of good sense, have given proofs of their capacity hitherto only when accompanied and supported by the former. They have not yet walked alone. They are taking their first steps in the dark. They are leading us, they say, to a regular, assured, measured, exactly-weighed state of well-being, to the conquest of matter. Be it so: they have charge of this kind of happiness. But let them not pretend that, to attain it, it is necessary to fling over-board, like a dangerous cargo, all that hitherto formed the heroic, cloud-topped, indefatigable, adventurous energy of our conscience. Leave us a few fancy virtues. Allow a little space for our fraternal senti-ments. It is very possible that these virtues and sentiments, which are not strictly in-dispensable to the just man of to-day, are the roots of all that will blossom when man shall have accomplished the hardest stage of the "struggle for life." Also, we must keep a few sumptuary virtues in reserve, in order to replace those which we abandon as useless; for our conscience has need of exercise and nourishment. Already we have thrown off a number of constraints which were assuredly hurtful, but which at least kept up the activity of our inner life. We are no longer chaste, since we have recognized that the work of the flesh, cursed for twenty centuries, is natural and lawful. We no longer go out in search of resignation, of mortification, of sacrifice; we are no longer lowly in heart or poor in spirit. All this is very lawful, seeing that these virtues depended on a religion which is retiring; but it is not well that their places should remain empty. Our ideal no longer asks to create saints, virgins, martyrs; but, even though it take another road, the spiritual road that animated the saints must remain intact and is still necessary to the man who wishes to go further than simple justice. It is beyond that simple justice that the morality begins of those who hope in the future. It is in this perhaps fairy-like, but not chimerical part of our conscience that we must acclimatize ourselves and learn to delight. It is still reasonable to persuade ourselves that in so doing we have not been duped.

22

The good-will of men is admirable. They are ready to renounce all the rights which they thought specific, to abandon all their dreams and all their hopes of happiness, even as many of them have already abandoned, without despairing, all their hopes beyond the tomb. They are resigned in advance to seeing their generations succeed one another without an object, a mission, an horizon, a future, if such be the certain will of life. The energy and pride of our conscience will manifest themselves for the last time in this acceptation and in this adhesion. But, before reaching this stage, before abdicating so gloomily, it is right that we should ask for proofs; and, hitherto, these seem to turn against those who bring them. In any case, nothing is decided. We are still in

Our Anxious Morality

suspense. Those who assure us that the old moral ideal must disappear, because the religions are disappearing, are strangely mistaken. It was not the religions that formed the ideal, but the ideal that gave birth to the religions. Now that these last have weakened or disappeared, their sources survive and seek another channel. When all is said, with the exception of certain factitious and parasitic virtues which we naturally abandon at the turn of the majority of religions, there is nothing as yet to be changed in our old Aryan ideal of justice, conscientiousness, courage, kindness and honour. We have only to draw nearer to it, to clasp it more closely, to realize it more effectively; and, before going beyond it, we have still a long and noble road to travel beneath the stars.

ROME

ROME

FOR twenty centuries, Rome has been the storehouse of all that was beautiful; and surely in no other spot in the world does so much beauty survive.

She has created nothing, save perhaps a certain spirit of grandeur, a co-ordination of beautiful things; but the most magnificent moments of the earth clung to her so fondly and displayed such energy during their sojourn that on no other point of the globe have they left so many imperishable traces. Treading her soil, we tread the mutilated footprint of the goddess who reveals herself no longer to men.

Nature gave her the wonderful site, established her fitly for the races that passed beside on the peaks of history to let fall their jewels into the noblest cup ever opened beneath the sky. She was not unworthy to receive those marvels ; she was already their equal. Beneath her limpid azure, the gloomy, obscure plants of the north still mate with southern foliage, inhaling their brightness and gladness. To the purest of her trees—the cypress, which lifts its head like an ardent and sombre prayer ; the stone-pine, into which the forest has whispered its gravest and sweetest thought; the massive evergreen oak, which so willingly adopts an archway's graceful form—to these the tradition of ages has given a pride, a conscious solemnity which they possess no elsewhere in the world. None can forget them, who once has seen them and understood, or fail to recognize them from among kindred trees of a less sacred soil. They were the ornaments, they were the witnesses of incomparable things. They are one with the scattered aqueducts, the discrowned mausoleums, the broken arches; one with the columns, heroic in their ruin, that array the deserted Campagna. They have assumed the style of the eternal marbles, which they

surround with respect and silence. Like
these marbles, they also have two or three
clear, but mysterious lines to tell of the
sorrow confessed by a plain that bears, without flinching, the wreck of its glory. They
are—and know they are—Roman.

A circle of mountains, their sonorous
names augustly familiar, their heads often
charged with snow as dazzling as the memories which they evoke, create around the
city that never can perish a precise and
glorious horizon, which divides her from
the world, but does not isolate her from the
sky. And, in these desolate precincts; in
the midst of the lifeless places where the
flagstones, the steps, the porticoes multiply
silence and absence; at all the cross-roads
where some wounded statue keeps guard in
emptiness; among the basins, the capitals,
the nymphs and the tritons, water flows
docile and luminous, obedient still to the
orders received two thousand years ago,
decking the immaculate solitude with its
mobile fragrance, its garlands of dew and
trophies of crystal, its azure plumes and
crowns of pearl. It is as though time,

among all the monuments that had hoped
to brave it, respected only the fragile hours
of that which evaporates and flows away.

2

Beauty, though always a borrowed beauty,
has dwelt so long within these walls which go
from the Janiculum to the Esquiline; it has
taken root there so persistently that the very
spot, the air we breathe, the sky that covers
it, the curves that define it have acquired a
prodigious power of appropriation and en-
noblement. Rome, like a pyre, purifies all
that the errors and caprices of men, their
ignorance and extravagance have forced
upon her incessantly since her ruin. So
far, it has been impossible to disfigure her.
One might almost believe that, for any work
to be carried out here or to live, it must
first cast off its original ugliness, it must
cease to be vulgar. Whatever does not
conform to the style of the seven hills is
slowly effaced and rejected; it crumbles
beneath the influence of the watchful genius
that has fixed the æsthetic principles of the

city on the horizons, the rocks and the marble of the heights. Thus, for instance, the art of the middle ages and the Primitives must have been more active here than in any other city, since this was the heart of the Christian universe; and yet they have left but few distinctive traces, these even appearing, as it were, hidden and ashamed: enough, but no more, for the history of the world, of which this was the centre, not to be left incomplete. But when we turn to those artists whose spirit was naturally in harmony with that which presides over the destinies of the eternal city—Giulio Romano, the Carracci and, above all, Raphael and Michael Angelo—we find in their work here a plenitude of power, a conviction, a kind of instinctive satisfaction which they manifest in no other place. One feels that they had not to create, but only to choose from among the unrevealed forms that thronged to them imperiously from every side, clamouring to be born: to these the masters gave substance. A mistake was impossible: they did not paint, in the proper sense of the word, but merely uncovered the veiled images which

Life and Flowers

had haunted the halls and arcades of the palaces. And so intimate, so indispensable is the relation between their art and the environment that gives it life that, when their works are exiled to the museums or churches of other cities, they seem out of proportion, unduly vigorous and unduly decorative, with an arbitrary conception of life. It is for this reason that copies or photographs of the ceiling of the Sistine Chapel appear disconcerting and almost incomprehensible.

But to the traveller who does not enter the Vatican till he, too, has drunk in the mighty will-power that emanates from the thousand fragments of the temples and public places: to him Michael Angelo's overpowering effort becomes magnificent and natural. The prodigious vault, on which a people of giants hurtle together in a grave and harmonious orgy of enthusiasm and muscles, turns into an arch of the very sky and reflects all the scenes of energy, all the burning virtues the memories of which are still restless beneath the ruins of this passionate soil. So, too, as he stands before

132

the *Conflagration of the Borgo*, he will not feel
as he would were he to behold the admirable
fresco on the walls of the National Gallery
or the Louvre ; he will not say to himself,
as Taine does, for instance, that these superb
nude bodies are but vaguely concerned with
the thing that is happening, that the flames
which arise from the building in no wise disturb them and that their one preoccupation
is to pose as good models and bring into
value the curve of a hip or the anatomy of
a thigh.   No, the visitor who has submissively heeded the injunctions of all that surrounds him will require no telling that here,
in these halls of the Vatican, as beneath the
vault of the Sistine, he is contemplating the
tardy, but normal and logical development of
an art which might have been that of Rome.
He will realize that, different as the impression may be which these two great efforts pro-
duce, he discovers the formula here which the
too positive genius of the Quirites had lacked
the good fortune or the opportunity to disengage.   For Rome, notwithstanding all her
endeavours, could not, of her own initiative,
give to the universe the essential image which

she had promised. It was to the spoils of
Greece that she owed her beauty; and her
chief merit was that she understood the
beauty of Greek art and eagerly amassed
its treasures. Her endeavours to add to it
resulted only in deformity; she was unable
to adapt its expression to her personal life.
Her paintings and sculptures responded only
by a kind of heresy, a vague approximate-
ness to the realities of her existence; and
such feeble originality as her architecture
possessed was due solely to its colossal proportions. One might almost imagine that old
Buonarotti and the superb colourist of Urbino
had but unearthed, after all the catas-
trophes, all the long silences and the seeming
deaths of Rome, the latent, uninterrupted
tradition which had unceasingly been in travail
underground and which now emerged at last
to culminate in their work and to declare
to the world what the Empire had been
powerless to say. For these men are more
distinctively Roman, more truly representative, perhaps, of the unconscious and secret
desire of that Latin earth than was the Rome
of the Cæsars. That Rome had failed in

its image. She had remained artificially
Hellenic; and Greece could not provide
this infinitely vaster race, differing so widely
from her, with the forms demanded by its
ornamental consciousness. Greece could be
only a sure and magnificent starting-point;
but her delicate, precise statues and paintings, so nicely, almost minutely proportioned,
were out of place in that Forum, surcharged
with immense monuments, as among the
monstrous Thermæ and violent circuses, or
under the sumptuous arches of the super-
posed basilicas. What if those frescoes of
Michael Angelo were the answer to the
call of the empty arches that had waited
a thousand years; what if they were the
almost organic consequence of those imperial
columns and marbles? And may we not
ask ourselves too whether the ceiling, the
pendentives and lunettes of the Farnesina
and the *Conflagration of the Borgo* do not
illustrate, better by far than the sculptures
of Phidias or Praxiteles, better also than the
best paintings of Pompeii or Herculaneum,
the *Metamorphoses* of Ovid, Virgil's *Æneid*,
or the poems of Horace?

Life and Flowers

3

But all this, perhaps, is merely illusion and due to the spell of the appropriative power which we have mentioned above. That power is such that whatever might, at the first glance, seem wholly opposed to the idea that reigns within these walls not only does not contradict this idea, but serves to define and declare it. Even Bernini—rhetorical, exuberant, ubiquitous Bernini—as irreconcilable as it is possible to be with the primitive gravity and taciturnity of Rome, even he, so detestable elsewhere, seems here to be adopted, justified by the genius of the city and serves to explain and illustrate certain somewhat redundant and declamatory sides of Roman greatness.

Moreover, a city that possesses the *Venus* of the Capitol and of the Vatican, the *Sleeping Ariadne*, the *Meleager* and the *Torso of Hercules*, the countless marvels of museums as numerous almost as her palaces—think only of the treasures in a single one of these museums, the newest of all, the Nazionale —a city whose every street, almost every

house conceals some fragment of marble or
bronze which, did some new town contain it,
would send pilgrims flocking; a city that
can offer the Pantheon of Agrippa, certain
columns in the Forum, in a word, so many
treasures that baffled memory cannot keep
pace with untiring admiration; a city that
has among its wonders those cypress-girdled
lawns of the Villa Borghese, those fountains,
those eternal gardens; a city, indeed, that
is the refuge of all that was best in the past
of the only people who cultivated beauty as
others cultivated corn, the olive or the vine :
such a city opposes a resistance to vulgarity
which, inactive though it be, is yet invincible ;
and she can tolerate all things without de-
filement. The immortal presence of an
assembly of gods, so perfect that no mutila-
tion can alter the rhythm of body or pose,
protects her against the errors herself may
commit and prevents the new generations of
men from having more empire upon her
than time and the barbarians had on those
very gods.

And these lead us back to the little cities of
Hellas that discovered one day and fixed for

Life and Flowers

ever the laws of human beauty. The beauty
of the earth, except for some spots which our
sordid industries have ravaged, has altered but
little since the days of Augustus and Pericles.
The sea is infinite still, is still inviolate. The
forest, the plain, the harvest, the villages,
rivers and streams, the mountains, the dawn
and the evening, the stars and the sky, vary
as these all may according to climate and
latitude, offer us still the same spectacles of
grandeur and tenderness, the same soft, profound harmonies, the same fairy-like scenes
of changing complexity which they showed
to the Athenian citizens and the people of
Rome. Nature remains more or less as she
was; and, besides, we have grown more
sensitive and can to-day admire more freely.
But, when we turn to the beauty special to
man, the beauty that is his own immediate
aim, we find that, owing perhaps to our too
great wealth or excessive application, to the
scattering of our efforts, our lack of concen-
tration, or the want of a certain goal and
an incontestable starting-point, we appear to
have lost almost all that the ancients had
been able to establish and make their own.

Rome

In all that regards purely human æsthetics, in
what concerns our body, our gestures, our
clothes, the objects we live with, our houses
and gardens, our monuments, even our land-
scapes, we are groping so timidly, we display
such confusion and inexperience that one
might truly believe our occupation of this
planet to date but from yesterday and ourselves to be still at the very beginning of the
period of adaptation. For the work of our
hands there no longer exists a common
measure, an accepted rule or conviction.
Our painters, our architects, our sculptors,
our men of letters—and we in our homes,
our cities—seek in a thousand different, con-
tradictory directions for the sure, the undeniable beauty which the ancients possessed so
fully. Should one of us, by any chance,
create, join together or discover a few lines,
a harmony of form or colour that should
incontestably prove that the mysterious, decisive point had been attained, it would be
regarded as the merest hazard, as an isolated
and precious phenomenon and neither the
author nor any one else would be able to
repeat it.

Life and Flowers

And yet, for a few happy years, man had mastered the laws of the beauty that is essentially and specifically human; and so great was his certainty that it compels our conviction even to this day. In the beauty of his own body, the Greek instinctively found the fixed standard which the Egyptians, the Assyrians, the Persians and all the anterior civilizations had sought in vain among animals and flowers, rocks and mountains, monsters and chimeras; and the architecture of his temples and palaces, the style of his houses, the proportion and ornament of the things which he used in his daily life were all derived from the beauty of this nude and perfect body. This people, among which nudity, with its natural consequence, the irreproachable harmony of limbs and muscles, was almost a religious and civic obligation, has taught us that the beauty of the human body is as diverse in its perfection, as spiritual, as mysterious as the beauty of the stars or sea. Every other ideal has misled and must always mislead the endeavours and efforts of man. In all the arts, intelligent races came nearer to true beauty in proportion

as they came nearer to the habit of nudity;
departing from this, they departed also
from beauty. The beauty proper to Rome
—in other words, the little original
beauty which she added to the spoils of
Greece—was due to the last remains of this
custom. For, in Rome, as Taine tells us,
" they also assembled together to swim, to
be rubbed, to perspire, to wrestle and run;
or at least, to watch the runners and wrest-
lers. For Rome, in this respect, is only an
enlarged Athens; the same ways of life
obtain, the same habits, the same instincts
and pleasures: the only difference lies in
the proportion and the moment. The city
has swollen till it numbers masters by the
hundred thousand and slaves by the million;
but, from Xenophon to Marcus Aurelius,
the gymnastic and rhetorical training has not
altered; they have still the tastes of athletes
and orators and it is in this direction that
one must work to please them; they are
worshippers of the nude, they are judges
of style, of conversation and ornament.
We can no longer understand this pagan
life of the body, which was so curious and

Life and Flowers

yet so idle; the climate has remained as it was, but man changed when he put on clothes and turned Christian."

It might more justly be said, perhaps, that Rome, at the period of which Taine speaks, was an intermittent and incomplete Athens. What was habitual there and, in some measure, organic becomes here only artificial and exceptional. They still cultivate and admire the human body, but it is almost always concealed by the toga; and the wearing of the toga blurs the pure, clear lines which a multitude of nude and living statues imposed upon the columns and pediments of the temples. The monuments grow larger and larger, lose their form and, little by little, their human harmony. The golden standard is shrouded and the veil shall be lifted only by a few artists of the Renascence, which was the moment when positive beauty shed its last beams.

THE PSYCHOLOGY OF ACCIDENT

THE PSYCHOLOGY OF ACCIDENT

I

THE more we master the forces of nature, the more do our chances of accidents multiply, even as the tamer's dangers increase in proportion to the number of wild animals which he "puts through their tricks" in the cage. Formerly, we avoided the contact of these forces as much as possible; to-day, they have gained admittance to our household. And so, notwithstanding our more prudent and peaceable manners, it happens to us more often than to our fathers to look pretty closely upon death. It is probable, therefore, that many of those who read these notes will have felt the same emotions and have had occasion to make similar remarks.

Life and Flowers

2

One of the first questions that arise is that of presentiment. Is it true, as many assert, that from the very morning we have a sort of intuition of the event that threatens the day? It is difficult to reply, inasmuch as our experience can bear only upon events which " might have turned out worse," or which, at least, have had no serious results. It seems natural, therefore, that those accidents which were to be free from consequences should not have stirred the deep waters of our instinct beforehand; and I believe it to be true that they do not even ripple their surface. As for the others, which entail a more or less speedy death, their victims seldom possess the strength or lucidity required to satisfy our curiosity. In any case, all that our personal experience is able to gather on this subject is very uncertain; and the question remains.

3

One fine day, then, we start at early dawn, by motor-car, bicycle, motor-cycle, in a skiff

The Psychology of Accident

or steam-boat : it is immaterial to the event that is preparing ; but, to make the picture more definite, let us take, by preference, a motor-car or motor-cycle, which are wonderful instruments of affliction and which put the fiercest questions to fortune in the great game of life and death. Suddenly, for no reason, at the turn of the road, in the very middle of the long, wide highway, at the top of a descent, here or there, on the right or on the left, seizing the brake, the wheel, the steering-handle, unexpectedly barring all space, assuming the deceptive and perfectly transparent appearance of a tree, a wall, a rock, an obstacle of one sort or another, stands death, face to face, towering, unforeseen, huge, immediate, indubitable, inevitable, irrevocable, and, with a click, shuts off the horizon of life, which it leaves without outlet. . . .

Forthwith, an eager and interminable scene, contained within half a second, sets in between our intelligence and our instinct. The attitude of our intelligence, our reason, our consciousness, by whatever name you please to call it, is extremely interesting. It decides instantaneously, sanely and logically that all is irretrievably lost. Yet it displays neither madness nor terror. It pictures the catastrophe, with all its details and consequences, exactly; and it realizes with contentment that it is not afraid and that it preserves its lucidity. Between the fall and the collision, it has time to rest, it reflects, it diverts itself, it finds leisure wherein to think of all manner of other things, to call up memories, to make comparisons, trifling and accurate observations: the tree which we see through death is a plane-tree, there are three holes in its patterned bark. . . . It is not so fine as the one in the garden. . . . The rock on which our skull will be broken is veined with mica and very white marble. . . . Our intelligence feels that it is not responsible, that we have nothing to reproach it with; it is almost smiling, it enjoys an ambiguous sensation of pleasure and awaits the inevitable with a tempered resignation mingled with prodigious curiosity.

The Psychology of Accident

It is evident that, if our lives had only the
intervention of this indolent, this too-logical
and too-clearsighted dilettante to rely upon,
every accident would be fated to end in
disaster. Luckily, warned by the nerves,
which whirl, lose their heads and bawl like
terrified children, another figure bounds
upon the stage, a rugged, brutal, naked,
muscular figure, elbowing its way and
seizing with an irresistible gesture such remnants of authority and chances of safety
as come within its reach. We call it instinct, the unconscious, the subconscious :
it matters not what we call it. Where was
it ? Where does it come from ? It was
somewhere asleep or else busied with dingy
and thankless tasks deep down in the primitive caverns of our body. Once it was that
body's uncontested king, but, for some
time since, has been relegated to the lower
darkness as an ill-bred, ill-dressed, ill-
spoken poor relation, a troublesome and
often disagreeeable witness of our original
misfortune. We no longer think of it,

no longer have recourse to it, save in the
desperate seconds of our supreme anguish.
Fortunately, it has a decent nature, is
utterly unselfish and bears no grudge.
Instinct knows, besides, that all those ornaments from the height of which we look
down upon and despise it are ephemeral and
frivolous and that, in reality, itself is the
sole master of the human dwelling. With
a glance that is surer and swifter than the
tremendous onrush of the peril, it takes in
the situation, then and there unravels all
its details, issues and possibilities and, in
a trice, affords a magnificent, an unforget-
table spectacle of strength, courage, precision and will, in which unconquered life
flies at the throat of unconquerable death.

5

This champion of existence, upstarting
like the shaggy savage of the fairy-tales
who comes to the rescue of the disconsolate
princess, works miracles in the strictest, the
most precise sense of the word. Above
all, under pressure of necessity, it has one

incomparable prerogative : it knows nothing of deliberation, of all the obstacles which it raises, all the impossibilities which it imposes. Instinct never accepts disaster, not for a moment admits the inevitable and, when on the point of being smashed to atoms, acts cheerfully against all hope, as though doubt, anxiety, fear, discouragement were notions absolutely foreign to the primitive forces that quicken it. Through a granite wall it sees nothing but safety, like a cranny of light ; and, by dint of seeing it, creates it in the stone. It does not abandon the hope of stopping a mountain that is rushing down upon it. It thrusts aside a rock, darts upon a wire, slips between two columns which were mathematically too close together to admit its passage. Among trees, it chooses infallibly the only one that will yield because an invisible worm has gnawed its root; amid a cluster of vain leaves, it discovers the one strong branch that overhangs the abyss; and, in a heap of sharp flints, it is as though instinct had prepared in anticipation the bed of moss and ferns that is to receive the body. . . .

The danger once past, reason, stupefied,

gasping for breath, unbelieving, a little disconcerted, turns its head to take a last look at the improbable. Then it resumes the lead, as of right, while the good savage, that no one dreams of thanking, returns in silence to its cave.

6

Perhaps it is not surprising that instinct should save us from the great habitual and immemorial dangers : water, fire, falls, collisions, animals. There is here, evidently, a long custom, an ancestral experience to explain its skill. But what amazes me is the ease, the quickness wherewith it acquaints itself with the most complicated, the most unusual inventions of our intelligence. We have only, once and for all, to show it the mechanism, the use and the purpose of the most unexpected machine, however foreign and useless to our real and primitive needs : instinct understands; and, from that moment, in an exigency, it will know the machine's last secrets and its management better than does the intelligence which constructed it.

The Psychology of Accident

That is why, let the instrument be as new, as recent or as formidable as it will, we can safely say that, in principle, there is no such thing as an inevitable catastrophe. Our unconsciousness is always alive and equal to every imaginable situation. Between the jaws of the vice contained in the power of the mountain or the sea, we can, we must look for a decisive movement on the part of our instinct, which possesses resources as inexhaustible as those of the universe or of nature, upon whose stores it draws at will.

7

And yet, if the whole truth be told, we no longer all have the same right to rely upon its sovereign intercession. It never dies, never sulks, is never mistaken ; but many men banish it to such depths, so rarely permit it to catch a glimpse of sunlight, lose sight of it so entirely, humiliate it so cruelly, pinion it so closely that, in the madness of their dire need, they forget where to look for it. They have not the material time in which to warn it or to release it from the

dungeon wherein they have chained it ; and,
when, at last, full of goodwill, armed with
its tools, it hurries up to the rescue, the
mischief is done, it is too late, death has
completed its work.

These inequalities of instinct, which are
connected rather, I suppose, with the prompt-
ness of the appeal than with the quality of
the assistance, appear in every accident.
Place two motorists in two parallel, includ-
able and exactly identical cases of danger :
an inexplicable touch of the wheel, a leap,
a twist, a turn, a sheer quiescence, a spell
of some kind will save the one, whereas
the other will go his normal and wretched
way and be smashed to pieces against the
obstacle. Of the six persons in a car, all
strictly involved in the same fate, three will
make the only possible, illogical, unforeseen
and necessary movement, while the three
others will act with too much intelligence
in the wrong direction. I once witnessed
one of those surprising manifestations of instinct, or nearly witnessed it ; for, although
I arrived after the accident, at least I gathered
the throbbing impressions on the spot, among

The Psychology of Accident

the injured. It was on the descent from
Gourdon, the rugged little village, well
known to excursionists from Cannes and
Nice, perched on a precipitous rock, over
two thousand feet in height, to escape the
Barbary pirates. It is inaccessible on every
side; no thoroughfare leads to it, save a
terrible zigzag way, which runs down between two ravines. A tilted cart, overloaded
with eight persons, including a woman carrying her child not two months old, was de-
scending this dangerous road, when the horse
took fright, ran away and darted towards the
abyss. The passengers felt themselves rushing to their deaths; and the woman, anxious
to save the child and obeying an admirable
impulse of maternal love, flung it, at the
supreme moment, from the other side of
the cart, where it fell on the roadway, while
all the others disappeared in the precipice
bristling with murderous rocks. Now, by
a miracle which is not unusual where human
lives are at stake, the seven victims, caught
up in brushwood, in all manner of boughs,
escaped with insignificant scratches, whereas
the poor little child died where it fell, with

its skull broken by a stone on the road. Two contrary instincts had here struggled for the mastery; and that one with which a glimmer of reflection had probably been mingled had made the more awkward movement of the two. You will speak of good and bad luck. These mysterious words are permissible, provided it be understood that they are applied to the mysterious movements of our unconsciousness. It is, in fact, preferable, whenever the thing is possible, to throw back the source of a mystery within ourselves: we thus limit to that extent the inauspicious field of error, discouragement and impotence.

8

We immediately ask ourselves whether we are able, if not to perfect our instinct, which I persist in believing perfect, at least to recall it closer to our will, to unloose its bonds, to restore its original freedom. This question would demand a special study. In the meantime, it appears fairly probable that, by drawing habitually and systematically

The Psychology of Accident

closer to material forces and facts, to all that which, in a word that expresses enormous things, we call nature, we can diminish by so much daily the distance which instinct will have to cover in order to come to our aid. This distance, as yet inappreciable in savages and in simple and humble men, increases with every step taken by our education and civilization. I am persuaded that it could be proved that a peasant or workman, even if he be the less young and the less active, if overtaken by the same disaster as his squire or employer, has two or three chances more than the latter of escaping safe and sound. In any case, there is no accident of which the victim is not, *a priori*, in the wrong. It is meet that he should say to himself, what is literally true, that any other, in his place, would have escaped ; consequently, the majority of the risks which those around him take remain forbidden to himself. His unconsciousness, which here blends with his future, is not " in form." Henceforth he must distrust his luck. From the point of view of the great dangers, he is a *minus habens*, as they used to say in Roman law.

Life and Flowers

9

For all this, when we consider the lack of consistency of our body, the inordinate power of all that surrounds it and the number of perils to which we expose ourselves, our luck, compared with that of other living beings, must needs appear prodigious. In the midst of our machines, our various apparatus, our poisons, our fires, our waters, all the forces which we have more or less mastered, but which are always ready to rise in revolt, we risk our lives twenty or thirty times oftener than the horse, for instance, the ox or the dog. Now, in a street or road accident, in a flood, an earthquake, a storm, a fire, in the fall of a tree or a house, the animal will almost always be struck by preference to the man. It is obvious that the latter's reason, his experience and his more prudent instinct preserve him to a great extent. Nevertheless, one would say that there must be something more. Granting equal risks and hazards and allowing for intelligence and a more skilful and certain instinct, the fact still remains that nature

The Psychology of Accident

seems to be afraid of man. She religiously
avoids touching that frail body, surrounds
it with a sort of manifest and unaccountable
respect and, when, through our own arrogant
fault, we oblige her to hurt us, she does us
the least harm possible.

IN PRAISE OF THE FIST

IN PRAISE OF THE FIST

I

IT is well, in the holiday season of summer, to occupy ourselves with the aptitudes of our body, once more restored to nature, and, in particular, with the exercises that most increase its strength, its agility and its qualities as the body of a fine animal, healthy, formidable, ready to face all life's exigencies.

I remember, in this connection, that lately, when writing of the sword,[1] I allowed myself to be carried away by my subject and was guilty of a certain injustice towards the only specific weapon with which nature has endowed us: I mean the fist. This injustice I am anxious to repair.

[1] *Cf.* the essay entitled *In Praise of the Sword*, in *The Double Garden*.—PUBLISHER'S NOTE.

Life and Flowers

The sword and the fist form each other's complement and, if the expression be not too ungracious, can keep house together on excellent terms. But the sword is, or should be, only an exceptional weapon, a sort of *ultima et sacra ratio*. We must not have recourse to it save with solemn precautions and a ceremonial equivalent to that wherewith we surround those criminal trials which may end in a sentence of death. The fist, on the contrary, is preeminently the everyday, the human weapon, the only weapon organically adapted to the sensibility, the resistance, the offensive and defensive structure of our body.

The fact is that, if we well examine ourselves, we must rank ourselves, without vanity, among the most unprotected, the most naked, the most fragile, the most brittle and flaccid beings in creation. Compare us, for instance, with the insect, so formidably equipped for attack and so fantastically armour-cased! Contemplate, among others, the ant, upon which you may heap ten or twenty thousand times the weight of its body without apparently

In Praise of the Fist

inconveniencing it. Consider the cockchafer, the least robust of the beetles, and weigh what it is able to carry before the rings of its abdomen crack or the casings of its fore-wings yield. As for the resistance of the stag-beetle, it is, so to speak, unlimited. By comparison, therefore, we and the majority of mammals are unsolidified beings, still in the gelatinous state and quite close to the primitive protoplasm. Our skeleton alone, which is as it were the rough sketch of our definitive form, offers a certain consistency. But how wretched is this skeleton, which one would think constructed by a child! Look at our spine, the basis of our whole system, whose ill-set vertebræ hold together only by a miracle, and our thoracic cage, which presents only a series of diagonals which we hardly dare touch with the finger-tips. Now it is against this slack and in-coherent machine, which resembles an abortive effort of nature, against this pitiful organism, from which life tends to escape on every side, that we have contrived weapons capable of annihilating us even if we possessed the fabulous armour-case, the prodigious strength and the incredible vitality of the most indestructible insects. We have here, it must be agreed, a very curious and a very disconcerting aberration, an initial folly, peculiar to the human race, that goes on increasing daily. In order to return to the natural logic followed by all other living beings, if we are permitted to use extraordinary weapons against our enemies of a different order, we ought, among ourselves, among men, to employ only the means of attack and defence provided by our own bodies. Were mankind to conform strictly to the evident will of nature, the fist, which is to man what its horns are to the bull and its claws and teeth to the lion, the fist should suffice for all our needs of protection, justice and revenge. A wiser race would forbid any other mode of combat as an irremissible crime against the essential laws of the species. At the end of a few generations, we should thus succeed in spreading and putting into force a sort of panic-stricken respect of human life. And how prompt and how exactly in accordance with nature's wishes would be the selection brought about by the intensive practice of pugilism, in which all the hopes of military glory would be centred. Now selection is, after all, the only really important thing that claims our preoccupation; it is the first, the greatest and the most eternal of our duties towards the race.

2

Meanwhile, the study of boxing gives us excellent lessons in humility and throws a somewhat alarming light upon the forfeiture of some of our most valuable instincts. We soon perceive that, in all that concerns the use of our limbs—agility, dexterity, muscular strength, resistance to pain—we have sunk to the lowest rank of the mammals or the batrachians. From this point of view, in a well-conceived hierarchy, we should be entitled to a modest place between the frog and the sheep. The kick of the horse, the butt of the bull, the bite of the dog are mechanically and anatomically perfect. It would be impossible to improve, by the most learned lessons, their instinctive manner of using their natural weapons. But we

Life and Flowers

the "Hominidæ," the proudest of the primates, do not know how to strike a blow with our fist! We do not even know which exactly is the weapon of our kind! Look at two draymen, two peasants who come to blows: nothing could be more pitiable. After a copious and dilatory broadside of insults and threats, they seize each other by the throat and hair, make play with their feet, with their knees, at random, bite each other, scratch each other, get entangled in their motionless rage, dare not leave go and, if one of them succeed in releasing an arm, he strikes out blindly and most often into space a series of hurried, stunted and sputtering little blows; and the combat would never end if the treacherous knife, evoked by the shame of the incongruous sight, did not suddenly, almost spontaneously leap from the pocket of one of the two.

On the other hand, watch two pugilists: no useless words, no gropings, no anger; the calmness of two certainties that know what lies before them. The athletic attitude of the guard, one of the finest of which

In Praise of the Fist

the male body is capable, logically exhibits all the muscles of the organism to the best advantage. From head to foot, not a particle of strength can now go astray. Every one of them has its pole in one or other of the two massive fists charged to the full with energy. And the noble simplicity of the attack! Three blows, no more, the fruits of secular experience, mathematically exhaust the thousand useless possibilities hazarded by the uninitiated. Three synthetic, irresistible, unimprovable blows. As soon as one of them frankly touches the adversary, the fight is ended, to the complete satisfaction of the conqueror, who triumphs so incontestably that he has no wish to abuse his victory, and with no dangerous hurt to the conquered, who is simply reduced to impotence and unconsciousness during the time needed for all illwill to evaporate. Soon after, the beaten man will rise to his feet with no lasting damage, because the resistance of his bones and his organs is strictly and naturally proportioned to the power of the human weapon that has struck him and brought him to the ground.

Life and Flowers

3

It may seem paradoxical, but the fact is easily established that the science of boxing, in those countries where it is generally practised and cultivated, becomes a pledge of peace and gentleness. Our aggressive nervousness, our watchful susceptibility, that sort of perpetual state of alarm in which our jealous vanity moves, all these arise, at bottom, from the sense of our weakness and of our physical inferiority, which toil as best they may to overawe, with a proud and irritable mask, the men, often churlish, unjust and malevolent, that surround us. The more that we feel ourselves disarmed in the face of attack, the more are we tortured by the longing to prove to others and to persuade ourselves that no one attacks us with impunity. Courage becomes the more touchy, the more intractable in proportion as our anxiously-terrified instinct, cowering within the body that is to receive the blows, asks itself how the bout will end. What will this poor prudent instinct do if the crisis goes badly? It is upon our instinct that we rely

in the hour of danger. Upon our instinct
devolve the anxiety of the attack, the care of
the defence. But we have so often in daily
life dismissed it from the control of affairs
and from the supreme council that, when
its name is called, it comes forth from its
retreat like one grown old in captivity and
suddenly dazzled by the light of day. What
resolution will it take? Where is it to
strike: at the eyes, the stomach, the nose,
the temples, the throat? And what weapon
is it to choose: the feet, the teeth, the
hand, the elbow, or the nails? It no longer
knows; it wanders about its poor dwelling,
which is about to be defaced; and, while, do-
tingly, it pulls them by the sleeve, courage,
pride, vanity, spirit, self-esteem, all the great
and splendid, but irresponsible lords envenom
the stubborn quarrel, which at last, after
numberless and grotesque evasions, ends in
an unskilful exchange of clamorous, blind,
ataxic thumps, hybrid and plaintive, piteous
and puerile and indefinitely impotent.

He, on the contrary, who knows the
source of justice which he holds in his two
closed fists has no need for self-persuasion.

Life and Flowers

Once and for all, he knows. Longanimity emanates like a peaceful flower from his ideal, but certain victory. The grossest insult cannot impair his indulgent smile. Peaceably he awaits the first act of violence and is able to say to all and any that offend him, "Thus far shall you go." A single magic movement stops the insolence. Why make this movement? He ceases even to think of it, so certain is its efficacy. And it is with a sense of shame, as of one striking a defenceless child, that, in the last extremity, he at length resolves to raise against the most powerful brute the sovereign hand that regrets beforehand its too-easy victory.

THE FORGIVENESS OF INJURIES

THE FORGIVENESS OF INJURIES

I⊤ is not unprofitable to examine from time
to time the meaning of certain words which
clothe in an unchangeable garment thoughts
that have themselves become transmuted.

To take for instance the word "forgive,"
which appears, at first sight, one of the
most beautiful in the language: does this
word still, did it ever possess the sense
of almost divine amnesty which we assign
to it? Is it not one of the terms that best
set forth the good-will of men, inasmuch
as it contains an ideal that has never been
realized? When we say to one who has
injured us, "I forgive you and all is forgotten," how much truth is there at the
bottom of this speech? At most, this,
which is the only engagement into which

we can enter: "I shall not try to harm you in my turn." The remainder, which we believe ourselves to be promising, does not depend upon our own will. It is impossible for us to forget the wrong that has been done us, because the profoundest of our instincts, that of self-preservation, has a direct interest in remembering it.

The man who, at a given moment, finds his way into our lives is never known to us as he is. For us he is only an image which he himself outlines in our memory. It is quite true that the life that animates him has an indefinable, but powerful self-revealing face. It conveys a host of promises, which are probably deeper and more sincere than the words or actions that will ere long belie them. But this great sign has little more than an ideal value. We are in a world wherein, either through force of circumstances or as the result of an initial error, very few beings live in accordance with the truth which their presence there foretells. At long last, our fretful experience teaches us to take no further account of this too mysterious face. A plain, hard mask covers it and bears the impress of all the acts and deeds that have affected us. Kindnesses illumine it with attractive and delicate colours, whereas offences channel it with deep grooves. In reality, it is only under this mask, modelled according to the recollection of pleasures or cares, that we perceive the man who approaches us; and to say to him, if he have offended us, that we forgive him is tantamount to telling him that we do not recognize him.

2

It is a question of knowing what influence this inevitable recognition will have upon our relations with the man who has injured us. In this, as in so many other respects, as soon as our good-will is roused, its first, as yet unconscious steps bring it back to the old road of the religious ideal. At the summit of this ideal, we might set up, as a symbol, the legendary group of the Christian woman burying, at the risk of her life, the execrated remains of Nero.

Life and Flowers

There is no denying that the action of this woman is greater and goes farther beyond human reason than the action of Antigone, which dominates pagan antiquity. Nevertheless, it does not exhaust the limits of Christian forgiveness. Suppose that Nero be not dead, but staggering on the last confines of life and that an heroic rescue alone can save him. The Christian will owe him this rescue, even though she know for certain that the life which she is restoring to him will, at the same time, bring back the persecution. She can rise higher still: imagine that she have to choose in the same moment of anguish between her brother and the enemy who will doom her to destruction. She will reach the topmost summit only by preferring the enemy.

3

Of this ideal, which is sublime even where an infinite reward for it is taken into account, what are we to think in a world that looks for nothing in another world?

The Forgiveness of Injuries

At which of the three superhuman moments shall we call him mad who flings himself into one of those three abysses of forgiveness? We shall even to this day find a few traces of footsteps around the first; but no one will now stray around the two others. Let us admit that we have here a sort of heroic march of faith which is no longer possible; but, taking away faith, there nevertheless remains, even in the unreason of that ideal, something human that is as it were a presentiment of what man would like to do if life were not so cruel.

And let us not think that instances of this kind, taken from the farthest ends of imagination, are idle or absurd. Existence constantly brings before us equivalents that are less tragic, but no less difficult; and the solution of the humblest cases of conscience depends upon the spirit which presides over that of the loftiest. All that we imagine on a large scale will end by being realized on a small; and upon the choice which we would make on the mountain depends exactly that which we will make in the valley.

Life and Flowers

4

Moreover, we can learn to forgive as completely as the Christian. We are no more prisoners than he of this world which we see with the eyes in our head. We need only an effort similar to his, but directed towards other gates, in order to escape from it. The Christian, just like ourselves, did not forget the injury; he did not attempt the impossible; but he first proceeded to drown any desire for revenge in the divine immensity. This divine immensity, more closely considered, is not very different from our own. Both, in reality, are but the feeling of the nameless immensity wherein we struggle. Religion raised every soul mechanically, so to speak, to the heights which we ought to reach by means of our own strength. But, as most of the souls which it drew thither were as yet blind, it made no vain endeavour to give them an idea of the truths which we perceive from those heights. They would not have understood them. It contented itself with describing to them pictures appropriate and familiar to their blindness, pictures which, for very different reasons, produced nearly the same effects as the real vision that strikes us at present. "We must forgive offences because God wishes it and has Himself set the most complete example of forgiveness that it is possible to imagine." This command, which we can follow without opening our eyes, is exactly the same as that given to us by the needs and the profound innocence of all life at the moment when we contemplate them from a sufficient height. And, if this latter command does not, like the first, go so far as to urge us to prefer our enemy because he is our enemy, this is not to say that it is less sublime, but that it addresses hearts which are more disinterested and minds which have learnt no longer to appraise an ideal solely according as to whether it be more or less difficult of attainment. In sacrifice, for instance, in penance, in mortification, there are, in this way, a whole series of spiritual victories which are more and more painful, but which are not really higher, because they rise not in the human atmosphere, but in the void above, where they shine not only without necessity, but often in a very hurtful fashion. The man who juggles with balls of fire on the point of a steeple is also doing a very difficult thing; yet no one dreams of comparing his useless courage with the devotion, nearly always less dangerous though it be, of the man who flings himself into the water or the flames to save a child. In any case—and perhaps more efficaciously than the other —the command of which we were speaking dispels all hatred, for it no longer springs from a foreign will, it is born within ourselves at the sight of an immense spectacle in which men's actions assume their real place and meaning. There is no more ill-will, ingratitude, injustice or perversity, there is not even any more selfishness, in the magnificent and boundless night wherein poor beings move, guided by a darkness which each of them follows in exceeding good faith, believing that he is fulfilling a duty or exercising a right.

The Forgiveness of Injuries

5

Let us not fear lest this vision, together with so many others which are grander and no less exact and which should always be present to our eyes, let us not fear lest it should disarm us and make victims or dupes of us in a life of vaster and harsher realities. There are very few among us that have need to strengthen their means of defence, to whet their prudence, their mistrust or their selfishness. Life's instinct and experience provide for this but too lavishly. We are never in danger of losing our equilibrium on the side opposed to our petty daily interests. All the efforts of a watchful thought suffice only with great difficulty to keep us erect. But it is no matter for indifference to others and especially to ourselves whether our movements of attack and defence are outlined against the dull background of hatred, contempt and disenchantment or against the transparent horizon of indulgence and of the silent forgiveness that explains and understands. Above all, as the years pass, let us keep to the humble lessons of experience. There is in these lessons a dull and heavy part that belongs by right to instinct and descends to the necessary clay-soil of life. There is no need to occupy ourselves with it: it buds and multiplies prodigiously in the unconscious. But there is a purer and more subtle part which we must learn to catch and hold before it evaporates in space. Every act allows of as many different interpretations as there are diverse forces in our intelligence. The lowest of them appear at first the simplest, the most natural and just, because they are the first to come, the idlest, those requiring the least effort. If we do not struggle without respite against their cunning and familiar encroachment, little by little they devour and poison all the hopes, all the beliefs out of which our youth had formed the noblest and most fruitful regions of our mind. Soon there would remain to us, towards the end of our days, nothing but the most miserable residue of wisdom. It is meet, therefore, that the loftiest interpretation which we can give of the facts that hustle us at every

The Forgiveness of Injuries

moment should rise in proportion as the gross treasure of the practical sense of existence accumulates. According as our sense of life increases in the soil by the roots, it is indispensable that it should ascend in the light by the fruits and flowers. It is necessary that an ever-vigilant thought should incessantly lift up, air and quicken the dead-weight of the years. Moreover, experience, seemingly so positive, so practical, so easy-going, so tranquil, so ingenuous and so sincere, knows full well that it hides some essential thing from us ; and, had we the strength to drive it to its most secret retrenchments, we should end to a certainty by wringing from it the supreme avowal that, upon the upshot and when all and everything is said, the loftiest interpretation is invariably the truest.

CONCERNING "KING LEAR"

CONCERNING "KING LEAR"

IT is easy to prove that, of late years and
especially since the beginning of the great
romantic period, the realm of poetry—which
had hardly been touched upon since the
definite loss of the vast, but uninhabitable
provinces of the epic poem—has gradually
shrunk in dimensions and become actually
reduced to a few isolated towns in the
mountain.   It will probably continue there,
long-lived and impregnable, and will gain
in purity and intensity all that it has lost
elsewhere in extent and abundance.   Little
by little it will strip itself of its vain didactic,
descriptive and narrative ornaments, soon
to be itself alone, that is to say the only
voice that can reveal to us the things which
silence hides from us, which human speech

no longer utters and which music does not yet express.

Lyric poetry will always exist: it is immortal, because it is necessary. But what fate has the future or even the present in store, I will not say for the dramatist or playwright, but for the tragic poet proper, for the writer who strives to maintain a certain lyrical quality in his work by representing in it things greater and finer than the things of real life?

It is certain that the lyric tragedy of the Greeks, that classical tragedy as conceived by Corneille and Racine, that the romantic tragedy of the Germans and Victor Hugo all derive their poetry from sources that are definitely dried up. The great drama of the crowds, in which it was believed that an unknown and inexhaustible source had been discovered, has hitherto yielded only mediocre and indifferent results. And the new mysteries of our modern life, which have taken the place of all the others and in the direction of which Ibsen attempted certain excavations, these mysteries have been for too short a time in direct contact with

man to erect and visibly and efficaciously
to govern the words and actions of the
character of a play. And yet there is no
disguising the fact and the poetic instinct of
humanity has always felt its presentiment : a
drama is not really true until it is greater
and finer than life.

2

Let us, in the interval preceding the time
when the poets shall know whither to turn
their steps, examine one of the most famous
examples of those dramas which enlarge the
truth without violating it, one of those rare
dramas which, after more than three centuries, still remain green and living in all
their parts : I allude to Shakspeare's *King
Lear*.

It is safe to declare, as I once said—
not without some little exaggeration, for
it is impossible to avoid exaggeration in
the light and exquisite attack of fever which
seizes all Shakspeare's devoted admirers
whenever one of his master-pieces is revived
—it is safe to declare, after surveying the

literatures of every period and of every
country, that the tragedy of the old king
constitutes the mightiest, the vastest, the
most stirring, the most intense dramatic
poem that has ever been written. Were
we to be asked from the height of another
planet which is the synthetic and representative play, the archetypal play of the human
stage, the play in which the ideal of the
loftiest scenic poetry is most fully realized, it
seems to me certain that, after due deliberation, all the poets of our earth, the best
judges in this exigency, would with one
voice name *King Lear*. They could only
for a moment weigh the claims of two or
three master-pieces of the Greek stage, or
else—for virtually Shakspeare can be compared with none save himself—of that other
miracle of his genius, the tragic story of
Hamlet Prince of Denmark.

3

Prometheus, the *Orestes*, *Œdipus Tyrannus*
are wonderful but isolated trees, whereas
King Lear is a marvellous forest. Let us

admit that Shakspeare's poem is less clear,
not so evident, not so visibly harmonious,
not so pure in outline, not so perfect in the
rather conventional sense of the word ; let
us grant that it has faults as enormous as
its good qualities : this fact none the less
remains, that it surpasses all the others in
the mass, the rarity, the density, the strange
mobility, the prodigious bulk of the tragic
beauties which it contains.   I know that the
total beauty of a work is not to be estimated
by weight or volume ; that the dimensions
of a statue do not necessarily bear a relation
to its æsthetic value.   Nevertheless, it cannot be denied that abundance, variety and
ampleness add certain vital, unaccustomed
elements to beauty ; that it is easier to be
successful with one statue of middling size
and of a calm movement than with a group
of twenty statues of superhuman dimensions,
endowed with passionate and yet coordinate
gestures ; that it is less difficult to write one
tragic and mighty act in which three or four
persons play their parts than to write five
which are filled with a whole moving crowd
and which maintain that same tragic and

powerful note on an equal level during a period five times as long as the other. Well, by the side of *King Lear*, the longest Greek tragedies are little more than plays in one act.

On the other hand, if we try to compare it with *Hamlet*, we shall probably find that its thought is less active, less acute, less profound, less quivering, less prophetic. By way of compensation, however, how much more vigorous, massive and irresistible does the spirit of the work appear! Certain clusters, certain rays of light on the platform of Elsinore reach and, for a moment, illumine, like gleams from beyond the tomb, more inaccessible darknesses; but here the column of smoke and flame lights up in a permanent and uniform manner a whole stretch of the night. The subject is simpler, more general and more normally human, the colouring more monotonous, but more majestically and more harmoniously superb, the intensity more constant and more widespread, the lyricism more continuous, more overflowing and more illusive and yet more natural, nearer to the realities of everyday life, more

194

familiarly stirring, because it springs not
from thought, but from passion, because it
surrounds a situation which, although exceptional, is, nevertheless, universally possible,
because it does not necessarily require a
metaphysical hero like Hamlet and because
it immediately affects the primitive and
almost invariable soul of man.

4

Hamlet, *Macbeth*, *Prometheus*, the *Orestes*,
Œdipus belong to a class of poems which are
more exalted than the others because they are
unfolded on a sort of sacred mountain girt
about by a certain mystery. This is what,
in the hierarchy of the master-pieces, places
Hamlet incontestably above *Othello*, for instance, although *Othello* is as passionately, as
profoundly and, doubtless, more normally
human. They owe to this mountain which
carries them between heaven and earth the
best part of their sombre and sublime power.
Now, if we examine the formation of this
mountain, we become aware that the elements
which compose it are borrowed from a

variable and arbitrary supernaturalism; it is
a "beyond" of a contestable character and
appearance, which are religious or supersti-
tious, transitory, therefore, or local. But—
and this it is that gives it a place apart
among the four or five great dramatic poems
of the world—in *King Lear* there is no
supernaturalism proper. The gods, the inhabitants of the great imaginary worlds do
not meddle with the action; fatality itself is
here quite inward, is no more than passion
run mad; and yet the immense drama unravels its five acts on a summit as high, as
overladen with spells, poetry and unwonted
anxieties as though all the traditional forces
of heaven and hell had vied in ardour to
superstruct its peaks. The absurdity of
the original anecdote (all the great master-
pieces, being intended to represent typical
actions of a necessarily far-fetched exclusive
and excessive character, are founded on a
more or less absurd anecdote) disappears in
the sublime magnificence of the height at
which it is developed. Study more closely
the structure of that summit: it is formed
solely of enormous human strata, of gigantic

blocks of passion, of reason, of general and almost familiar sentiments, overthrown, heaped up, superimposed by an awful tempest, but one profoundly suited to all that is most human in human nature.

That is why *King Lear* remains the youngest of the great tragic works, the only one which time has not withered. It needs an effort of our good-will, a forgetting of our condition and of our present certainties for us to be sincerely and wholly stirred by the spectacle of *Hamlet*, *Macbeth* or *Œdipus*. On the other hand, the wrath, the roars of pain, the prodigious curses of the old man, of the outraged father seem to issue from our modern hearts and brains; they rise up under our own sky; and, in respect of all the profound truths that form the spiritual and sentimental atmosphere of our planet, there is nothing essential to be added to them, nothing to be withdrawn from them. Were Shakspeare to return among us upon earth, he could no longer write *Hamlet* or *Macbeth*. He would feel that the main august and gloomy ideas upon which those poems rest would no longer carry them, whereas he would not

Life and Flowers

need to alter a situation nor a line in *King Lear*.

5

The youngest, the most unchangeable of tragedies is also the most organically lyrical dramatic poem that was ever realized, the only one in the world in which the magnificence of the language does not once impair the probability, the naturalness of the dialogue. There is not a poet but knows that it is almost impossible on the stage to ally beautiful images with natural expression. There is no denying it: no scene in the mightiest tragedy or in the most hackneyed comedy, as Alfred de Vigny said, is ever more than a conversation between two or three people who have met to talk of their affairs. They have therefore to talk; and, in order to give us that which is the most necessary illusion on the stage, the illusion of reality, they must depart as little as possible from the language employed in everyday life. But, in this rather elementary life, we hardly ever express in words anything that is brilliant or profound in our inner existence. If our habitual thoughts

mingle with great and beautiful spectacles,
with the highest mysteries of nature, they
remain within ourselves in a latent condition,
in a condition of dreams, of ideas, of mute
feelings which, at the very most, betray
themselves sometimes by a word, a phrase
nobler or truer than those of our probable
and usual conversation. Now, the drama
being able to express hardly anything that
would not be expressed in life, it follows
that all the higher part of existence remains
unformulated there, lest it should shatter
the indispensable illusion. The poet has
therefore to choose : he will be lyrical or
merely eloquent, but unreal (and this is
the mistake of our classical tragedies, of
the plays of Victor Hugo and of almost
all the French and German romanticists, a
few scenes of Goethe excepted), or else he
will be natural, but dry, prosaic and dull.
Shakspeare did not escape the dangers of this
choice. In *Romeo and Juliet*, for instance,
and in most of his historical plays, he pours
forth into rhetoric and incessantly sacrifices
to the splendour, to the abundance of his
metaphors the imperious, essential precision

and commonplace of every speech and
cue.

6

On the other hand, in his great master-
pieces he makes no mistake ; but the very
manner in which he surmounts the difficulty reveals all the gravity of the problem. He achieves his end only with the
aid of a sort of subterfuge to which he
always resorts. As it seems to be accepted
that a hero who expresses his inner life in
all its magnificence cannot remain probable
and human on the stage except on condition that he be represented as mad in real
life (for it is understood that here the mad
alone express that hidden life), Shakspeare
systematically unsettles the reason of his pro-
tagonists and thus opens the dike that held
captive the swollen lyrical flood. Henceforward, he speaks freely by their mouths ;
and beauty invades the stage without fear-
ing lest it be told that it is out of place.
Henceforward, also, the lyricism of his great
works is more or less high, more or less wide
in proportion to the madness of his hero.

Concerning "King Lear"

Thus it is intermittent and restrained in *Macbeth* and *Othello*, because the hallucinations of the Thane of Cawdor and the rages of the Moor of Venice are no more than passional crises; it is slow and pensive in *Hamlet*, because the madness of the Prince of Denmark is torpid and meditative; but no otherwhere does it overflow as in *King Lear*, torrential, uninterrupted and irresistible, hurling together, in immense and miraculous images, the oceans, the forests, the tempests and the stars, because the magnificent insanity of the dispossessed and desperate old king extends from the first scene to the very last.

THE INTELLIGENCE OF THE FLOWERS

THE INTELLIGENCE OF THE FLOWERS

I

I WISH merely to recall here a few facts known to every botanist. I have made not a single discovery and my modest contribution is confined to a few elementary observations. I need hardly say that I have no intention of reviewing all the proofs of intelligence which the plants give us. These proofs are innumerable and continual, especially among the flowers, in which the effort of vegetable life towards light and understanding is concentrated.

Though there be plants and flowers that are awkward or unlucky, there is none that is wholly devoid of wisdom and ingenuity. All exert themselves to accomplish their work, all have the magnificent ambition to

overrun and conquer the surface of the
globe by endlessly multiplying that form of
existence which they represent. To attain
this object, they have, because of the law
that chains them to the soil, to overcome
difficulties much greater than those opposed
to the increase of the animals. And therefore the majority of them have recourse to
combinations, to a machinery, to traps which,
in regard to such matters as mechanism,
ballistics, aerial navigation and the observation of insects, have often anticipated the
inventions and acquirements of man.

2

It would be superfluous once more to
trace the picture of the great systems of
floral fertilization: the play of stamens and
pistil, the seduction of perfumes, the appeal
of harmonious and dazzling colours, the
concoction of nectar, which is absolutely
useless to the flower and is manufactured
only to attract and retain the liberator
from without, the messenger of love—bee,
humble-bee, fly, butterfly or moth—that is

The Intelligence of the Flowers

to bring to the flower the kiss of the distant, invisible, motionless lover. . . .

This vegetable world, which to us appears so placid, so resigned, in which all seems acquiescence, silence, obedience, meditation, is, on the contrary, that in which impatience, the revolt against destiny are the most vehement and stubborn. The essential organ, the nutrient organ of the plant, its root, attaches it indissolubly to the soil. If it be difficult to discover among the great laws that oppress us that which weighs heaviest upon our shoulders, in the case of the plant there is no doubt: it is the law that condemns it to immobility from its birth to its death. Therefore it knows better than we, who disseminate our efforts, against what first to rise in revolt. And the energy of its fixed idea, mounting from the darkness of the roots to become organized and full-blown in the flower, is an incomparable spectacle. It exerts itself wholly with one sole aim: to escape above from the fatality below, to evade, to transgress the heavy and sombre law, to set itself free, to shatter the narrow sphere, to

Life and Flowers

invent or invoke wings, to escape as far as
it can, to conquer the space in which destiny
encloses it, to approach another kingdom,
to penetrate into a moving and active world.
. . . . Is the fact that it attains its object not
as surprising as though we were to succeed
in living outside the time which a different
destiny assigns to us or in making our way
into a universe freed from the weightiest
laws of matter? We shall see that the
flower sets man a prodigious example of
insubmission, courage, perseverance and ingenuity. If we had applied to the removal
of various necessities that crush us, such as
pain, old age and death, one half of the
energy displayed by any little flower in our
gardens, we may well believe that our lot
would be very different from what it is.

3

This need of movement, this craving for
space, among the greater number of plants,
is manifested in both the flower and the
fruit. It is easily explained in the fruit, or,
in any case, discloses only a less complex

experience and foresight. Contrary to that which takes place in the animal kingdom and because of the terrible law of absolute immobility, the chief and worst enemy of the seed is the paternal stock. We are in a strange world, where the parents, unable to move from place to place, know that they are condemned to starve or stifle their offspring. Every seed that falls at the foot of the tree or plant is either lost or doomed to sprout in wretchedness. Hence the immense effort to throw off the yoke and conquer space. Hence the marvellous systems of dissemination, of propulsion, of navigation of the air which we find on every side in the forest and the plain: among others, to mention, in passing, but a few of the most curious, the aerial screw or samara of the Maple; the bract of the Lime-tree; the flying-machine of the Thistle, the Dandelion and the Salsafy; the detonating springs of the Spurge; the extraordinary squirt of the Momordica; the hooks of the eriophilous plants; and a thousand other unexpected and astounding pieces of mechanism; for there is not, so to speak, a single seed but has invented for its sole use a complete method of escaping from the maternal shade.

It would, in fact, be impossible, if one had not practised a little botany, to believe the expenditure of imagination and genius in all the verdure that gladdens our eyes. Consider, for instance, the charming seed-pots of the Scarlet Pimpernel, the five valves of the Balsam, the five bursting capsules of the Geranium. Do not forget, upon occasion, to examine the common Poppy-head, which we find at any herbalist's. This good, big head shelters a prudence and a foresight that deserve the highest praise. We know that it holds thousands of tiny black seeds. Its object is to scatter this seed as dexterously and to as great a distance as possible. If the capsule containing it were to split, to fall or to open underneath, the precious black dust would form but a useless heap at the foot of the maternal stalk. But its only outlet is through apertures contrived right at the top of the capsule, which, when ripe, bends over on its peduncle, sways like a censer at the least breath of wind and literally sows the seeds in space, with the very action employed by the sower.

Shall I speak of the seeds which provide for their dissemination by birds and which, to entice them, as in the case of the Mistletoe, the Juniper, the Mountain-ash, lurk inside a sweet husk? We see here displayed such a powerful reasoning faculty, such a remarkable understanding of final causes that we hardly dare dwell upon the subject, for fear of repeating the ingenuous mistakes of Bernardin de Saint-Pierre. And yet the facts can be no otherwise explained. The sweet husk is of no more use to the seed than the nectar, which attracts the bees, is to the flower. The bird eats the fruit because it is sweet and, at the same time, swallows the seed, which is indigestible. He flies away and, soon after, ejects the seed in the same condition in which he has received it, but stripped of its case and ready to sprout far from the attendant dangers of its birthplace.

Life and Flowers

4

But let us return to simpler contrivances.
Pick a blade of grass by the roadside, from
the first tuft that offers, and you will
perceive an independent, indefatigable, unexpected little intelligence at work. Here,
for instance, are two poor creeping plants
which you have met a thousand times on
your walks, for we find them in every spot,
down to the most ungrateful corners to
which a pinch of soil has strayed. They
are two varieties of wild Lucern or Medick
(*Medicago*), two "ill weeds" in the
humblest sense of the word. One bears a
reddish flower, the other a little yellow ball
the size of a pea. To see them crawling
and hiding among the proud grasses, one
would never suspect that, long before the
illustrious geometrician and physician of
Syracuse, they had discovered the Archimedean screw and endeavoured to apply
it not to the raising of liquids, but to the
art of flying. They lodge their seeds in
light spirals with three or four convolu-
tions, admirably constructed to delay their

fall and, consequently, with the help of the wind, to prolong their journey through the air. One of them, the yellow, has even improved upon the apparatus of the red by furnishing the edges of the spiral with a double row of points, with the evident intention of hooking it, on its passage, to either the clothes of the pedestrians or the fleece of the animals. It clearly hopes to add the advantages of eriophily — that is to say the dissemination of seed by sheep, goats, rabbits and so on—to those of anemophily, or dissemination by the wind.

The most touching side of this great effort is its futility. The poor red and yellow Lucerns have blundered. Their remarkable screws are of no use to them: they could act only if they fell from a certain height, from the top of some lofty tree or tall *Graminea ;* but, constructed as they are on the level of the grass, they have hardly taken a quarter of a turn before already they touch the ground. We have here a curious instance of the mistakes, the gropings, the experiments and the frequent little miscalculations of nature; for only those who have studied nature but very little will maintain that she never errs.

Let us observe, in passing, that other varieties of the Lucern (not to speak of the Clover, another papilionaceous *Leguminosa*, almost identical with that of which we are now speaking) have not adopted this flying apparatus, but keep to the primitive methods of the pod. In one of them, the *Medicago aurantiaca*, we very clearly perceive the transition from the twisted pod to the screw or spiral. Another variety, the *Medicago scutellata*, or Snail-medick, rounds its screw in the form of a ball. It would seem, therefore, that we are assisting at the stimulating spectacle of a sort of work of invention, at the attempts of a family that has not yet settled its destiny and is seeking for the best way of ensuring its future. Was it not perhaps, in the course of this search that, having been deceived in the spiral, the yellow Lucern added points or hooks to it, saying to itself, not unreasonably, that, since its leaves attract the sheep, it is inevitable and right that the sheep should assume the care of its progeny?

And, lastly, is it not thanks to this new effort and to this happy thought that the Lucern with the yellow flowers is infinitely more widely distributed than its sturdier cousin whose flowers are red?

5

It is not only in the seed or the flower, but in the whole plant, leaves, stalks and roots, that we discover, if we stoop for a moment over their humble work, many traces of a prudent and quick intelligence. Think of the magnificent struggle towards the light of the thwarted branches, or the ingenious and courageous strife of trees in danger. As for myself, I shall never forget the admirable example of heroism given me the other day in Provence, in the wild and delightful gorges of the Loup, all fragrant with violets, by a huge centenarian Laurel-tree. It was easy to read on its twisted and, so to speak, writhing trunk the whole drama of its hard and tenacious life. A bird or the wind, masters of destiny both, had carried the seed to the flank of the rock, which was as perpendicular as an iron curtain; and the tree was born there, two hundred yards above the torrent, inaccessible and solitary, among the burning and barren stones. From the first hour, it had sent its blind roots on a long and painful search for precarious water and soil. But this was only the hereditary care of a species that knows the aridity of the South. The young stem had to solve a much graver and more unexpected problem: it started from a vertical plane, so that its top, instead of rising towards the sky, bent down over the gulf. It was obliged, therefore, notwithstanding the increasing weight of its branches, to correct the first flight, stubbornly to bend its disconcerted trunk in the form of an elbow close to the rock and thus, like a swimmer who throws back his head, by means of an incessant will, tension and contraction to hold its heavy crown of leaves straight up into the sky.

Thenceforward, all the preoccupations, all the energy, all the free and conscious genius of the plant had centred around that vital knot. The monstrous, hypertrophied elbow

The Intelligence of the Flowers

revealed, one by one, the successive solicitudes of a kind of thought that knew how to profit by the warnings which it received from the rains and the storms. Year by year, the leafy dome grew heavier, with no other care than to spread itself out in the light and heat, while a hidden canker gnawed deep into the tragic arm that supported it in space. Then, obeying I know not what order of the instinct, two stout roots, two fibrous cables, issuing from the trunk at more than two feet above the elbow, had come to moor it to the granite wall. Had they really been evoked by the tree's distress or were they perhaps waiting providently, from the first day, for the acute hour of danger, in order to increase the value of their assistance? Was it only a happy accident? What human eye will ever assist at these silent dramas, which are all too long for our short lives?[1]

[1] Let us compare with this the act of intelligence of another root, whose exploits are related by Brandis in his *Ueber Leben und Polaritat.* In penetrating into the earth, it had come upon an old boot-sole : in order to cross this obstacle, which, apparently, the root was the first of its kind to find upon its road, it subdivided itself into as

Life and Flowers

6

Among the vegetals that give the most striking proofs of intelligence and initiative, the plants which might be described as "animated" or "sentient" deserve to be studied in detail. I will do no more than recall the delightful nervous terrors of the Sensitive-plant, the shrinking Mimosa with which we are all acquainted. There are other herbs endowed with spontaneous movements that are not so well known, notably the *Hedysareæ*, among which the *Hedysarum gyrans*, or Moving-plant, acts in a very restless and surprising fashion. This little *Leguminosa*, which is a native of Bengal, but often cultivated in our hothouses, performs a sort of perpetual and intricate dance in honour of the light. Its leaves are divided into three folioles, one wide and terminal, the two others narrow and planted at the base of the first. Each of these leaflets is

many parts as there were holes left in the sole by the stitching-needle ; then, when the obstacle was overcome, it came together again and united all its divided radicles into a single homogeneous tap-root.

animated with a different movement of its
own. They live in a state of rhythmical,
almost chronometrical and continuous agi-
tation. They are so sensitive to light
that their dance flags or quickens according as the clouds veil or uncover that
corner of the sky which they contemplate.
They are, as we see, real photometers; and
this long before Crook's discovery of the
natural otheoscopes.

7

But these plants, to which should be
added the Droseras, the Dionæas and many
others, are nervous plants that already go
a little beyond the mysterious and probably
imaginary ridge that separates the vegetable
from the animal kingdom. It is not necessary to seek so high; and we find as much
intelligence and almost as much visible
spontaneity at the other end of the world
which we are considering, in the low-lying
places where the plant is hardly distinct
from clay or stone. We have here the fabulous class of the *Cryptogamia*, which can be

studied only under the microscope, for which
reason we will pass it by in silence, although
the work of the sporules of the Mushrooms,
Ferns and Horse-tails is incomparable in its
delicacy and ingenuity. But, among the
aquatic plants, the inhabitants of the original ooze and mud, we can see less secret
marvels performed. As the fertilization
of their flowers cannot be accomplished
under water, each of them has thought
out a different system to allow of the dry
dissemination of the pollen. Thus, the
Zosteras, that is to say, the common Sea-
wrack with which we stuff our beds, carefully enclose their flower in a regular
diving-bell; and the Water-lilies send theirs
to blossom on the surface of the pond,
supporting and feeding it at the top of an
endless stalk, which lengthens as the level
of the water rises. The *Villersia nymphoides*,
having no expanding stalk, simply lets its
flowers go: they rise to the surface and
burst like bubbles. The *Trapa natans*, or
Water-caltrop, supplies them with a sort of
inflated tumour: they shoot up and open.
Then, when the fertilization is accomplished,

the air in the tumour is replaced by a muci-
laginous fluid, heavier than the water, and
the whole apparatus sinks back again to
the slime, where the fruits ripen.

The system of the Utricularia is even
more complicated. M. Henri Bocquillon
describes it in his *Vie des Plantes*:

"These plants, which are common in
ponds, ditches, pools and the puddles of
peat-bogs, are not visible in winter, when
they lie on the mud. Their long, slim,
trailing stalk is furnished with leaves reduced to ramified filaments. At the axilla
of the leaves thus transformed, we see a sort
of little pyriform pocket with an aperture in
its pointed upper end. This aperture has a
valve, which can be opened only from the
outside inwards; its edges are provided with
ramified hairs; the inside of the pocket
is covered with other little secretory hairs
which give it the appearance of velvet.
When the moment of efforescence has come,
the axillary utricles fill with air: the more
this air tends to escape, the more tightly
it closes the valve. The result is that it

Life and Flowers

imparts a great specific buoyancy to the plant
and carries it to the surface of the water.
Not till then do those charming little yellow
flowers come into blossom, resembling quaint
little mouths with more or less swollen lips
and palates streaked with orange or rubigi-
nous lines. During the months of June, July
and August, they display their fresh colours
gracefully above the muddy water, amid
the vegetable decay around them. But
fertilization has been effected, the fruit
develops, all things play a different part:
the ambient water presses upon the valve
of the utricles, forces it in, rushes into the
cavity, weighs down the plant and compels
it to descend to the mud again."

Is it not interesting to see thus gathered in
this immemorial little apparatus some of the
most fruitful and recent of human inventions:
the play of valves or plugs, the pressure of
fluids and air, the Archimedean principle
studied and turned to account? As the
author whom we have just quoted observes,
"The engineer who first attached a rafting
apparatus to a sunk ship little thought

that a similar process had been in use for thousands of years." In a world which we believe unconscious and destitute of intelligence, we begin by imagining that the least of our ideas creates new combinations and relations. When we come to look into things more closely, it appears infinitely probable that it is impossible for us to create anything whatsoever. We are the last comers on this earth, we simply find what has always existed and, like astonished children, we travel again the road which life had travelled before us. When all is said, it is very natural and comforting that this should be so. But we will return to this point.

8

We must not leave the aquatic plants without briefly mentioning the life of the most romantic of them all: the legendary Vallisneria, an hydrocharad whose nuptials form the most tragic episode in the love-history of the flowers. The Vallisneria is a rather insignificant herb, possessing none of the strange grace of the Water-lily or of certain submersed verdant plants. But it seems as though nature had delighted in giving it a beautiful idea. Its whole existence is spent at the bottom of the water, in a sort of half-slumber, until the moment of the wedding-hour comes, when it aspires to a new life. Then the female plant slowly uncoils the long spiral of its peduncle, rises, emerges and floats and blossoms on the surface of the pond. From a neighbouring stem, the male flowers, which see it through the sunlit water, rise in their turn, full of hope, towards the one that rocks, that awaits them, that calls them to a fairer world. But, when they have come half-way, they feel themselves suddenly held back: their stalk, the very source of their life, is too short; they will never reach the abode of light, the only spot in which the union of the stamens and the pistil can be achieved! . . .

Is there any more cruel inadvertence or ordeal in nature? Picture the tragedy of that longing, the inaccessible so nearly attained, the transparent fatality, the impossible with not a visible obstacle! It would be insoluble, like our own tragedy upon this earth, were it not that an unexpected element is mingled with it. Did the males foresee the disillusion to which they would be subjected? One thing is certain, that they have locked up in their hearts a bubble of air, even as we lock up in our souls a thought of desperate deliverance. It is as though they hesitated for a moment; then, with a magnificent effort, the finest, the most supernatural that I know of in all the pageantry of the insects and the flowers, in order to rise to happiness they deliberately break the bond that attaches them to life. They tear themselves from their peduncle and, with an incomparable flight, amid bubbles of gladness, their petals dart up and break the surface of the water. Wounded to death, but radiant and free, they float for a moment beside their heedless brides and the union is accomplished, whereupon the victims drift away to perish, while the wife, already a mother, closes her corolla, in which lives their last breath, rolls up her spiral and descends to the depths, there to ripen the fruit of the heroic kiss.

Must we spoil this charming picture, which is strictly accurate, but seen from the side of the light, by looking at it also from that of the shadow? Why not? There are sometimes on the shady side truths quite as interesting as those on the bright. This delightful tragedy is perfect only when we consider the intelligence and the aspirations of the species. But, when we observe individuals, we shall often see them act awkwardly and in the wrong way in this ideal plan. At one time, the male flowers will ascend to the surface when there are not yet any pistilled flowers near. At another, when the low water would permit them easily to join their companions, they will nevertheless mechanically and to no purpose break their stalks. We here once more establish the fact that all genius lies in the species, in life or in nature, whereas the individual is nearly always stupid. In man alone does a real emulation exist between the two intelligences, a more and more precise, more and more active tendency towards a sort of equilibrium which is the great secret of our future.

9

The parasitic plants, again, present curious and crafty spectacles, as in the case of the astonishing Cuscuta, commonly called the Dodder. It has no leaves; and no sooner has its stalk attained a few inches in length than it voluntarily abandons its roots to twine about its chosen victim, into which it digs its suckers. Thenceforth, it lives exclusively upon its prey. Its perspicacity is not to be deceived; it will refuse any support that does not please it and will go some distance, if necessary, in search of the stem of Hemp, Hop, Lucern or Flax that suits its temperament and its taste.

This Cuscuta naturally calls our attention to the Creepers, which have very remarkable habits and which deserve a word to themselves. Those of us who have lived a little in the country have often had occasion to admire the instinct, the sort of power of vision, that directs the tendrils of the Virginian Creeper or the Convolvulus towards the handle of a rake or spade resting against a wall. Move the rake and, the next day, the tendril will have turned completely round and found it again. Schopenhauer, in his treatise *Ueber den Willen in der Natur*, in the chapter devoted to the physiology of plants, recapitulates on this point and on many others a host of observations and experiments which it would take too long to set out here. I therefore refer the reader to this chapter, where he will find numerous sources and references marked out for him. Need I add that, in the past sixty or seventy years, these sources have been strangely multiplied and that, besides, the subject is almost inexhaustible?

Among so many different inventions, artifices and precautions, let us mention also, for instance, the foresight displayed by the *Hyoseris radiata*, or Starry Swine's-succory, a little yellow-flowered plant, not unlike the Dandelion and often found on the walls of the Riviera. In order to ensure both the dissemination and the stability of its race, it bears at one and the same time two kinds of seeds : the first are easily detached and are furnished with wings wherewith to abandon themselves to the wind, while the others have no wings, remain captive in the inflorescence and are set free only when the latter is decomposed.

The case of the *Xanthium spinosum*, or Spiny Xanthium, shows us how well-conceived and effective certain systems of dissemination can be. This Xanthium is a hideous weed, bristling with barbaric prickles. Not long ago, it was unknown in Western Europe and no one, naturally, had dreamt of acclimatizing it. It owes its conquests to the hooks which finish off the capsules of its fruits and which cling to the fleece of the animals. A native of Russia, it came to us in bales of wool imported from the depths of the Muscovite steppes; and one might follow on the map the stages of this great emigrant which has annexed a new world.

The *Silene Italica*, or Italian Catchfly, a simple little white flower, found in abundance under the olive-trees, has set its thought working in another direction. Apparently very timorous, very susceptible, to

avoid the visits of importunate and in-
delicate insects it furnishes its stalks with
glandular hairs, whence oozes a viscid fluid
in which the parasites are caught with
such success that the peasants of the South
use the plant as a fly-catcher in their
houses. Certain kinds of Catchflies, moreover, have ingeniously simplified the system.
Dreading the ants in particular, they discovered that it was enough, in order to
prevent them from passing, to place a wide
viscid ring under the node of each stalk.
This is exactly what our gardeners do
when they draw a circle of tar around
the trunk of the apple-trees to stop the
ascent of the caterpillars.

This leads to the study of the defensive
means employed by the plants. In an excellent popular work, *Les Plantes originales*,
to which I refer the reader who wishes for
fuller details, M. Henri Coupin examines
some of these quaint and startling weapons.
We have first the stimulating question of
the thorns, concerning which M. Lothelier,
a student at the Sorbonne, has made a
number of interesting experiments, resulting

in the conclusion that shade and damp tend to suppress the prickly parts of the plants. On the other hand, whenever the place in which it grows is dry and burnt by the sun, the plant bristles and multiplies its spikes, as though it felt that, as almost the sole survivor among the rocks or in the hot sand, it is called upon to make a mighty effort to redouble its defences against an enemy that no longer has a choice of victims to prey upon. It is a remarkable fact, moreover, that, when cultivated by man, most of the thorny plants gradually lay aside their weapons, leaving the care of their safety to the supernatural protector who has adopted them in his fenced grounds.[1]

[1] Among the plants that have ceased to defend themselves, the most striking case is that of the Lettuce :

"In its wild state," says the author whom I have mentioned above, "if we break a stalk or a leaf, we see a white juice exude from it, the *latex*, a substance formed of different matters which vigorously defend the plant against the assaults of the slugs. On the other hand, in the cultivated species derived from the former, the *latex* is almost missing, for which reason the plant, to the despair of the gardeners, is no longer able to resist and allows the slugs to eat it."

It is nevertheless right to add that this *latex* is rarely

Certain plants, among others the *Boragineæ*, supply the place of thorns with very hard bristles. Others, such as the Nettle, add poison. Others, the Geranium, the Mint, the Rue, steep themselves in powerful odours to keep off the animals. But the strangest are those which defend themselves mechanically. I will mention only the Horse-tail, which surrounds itself with a veritable armour of microscopic *silicæ*. Moreover, almost all the *Gramineæ*, in order to discourage the gluttony of the slugs and snails, add lime to their tissues.

10

Before broaching the study of the complicated forms of apparatus rendered necessary by cross-fertilization, among the thousands

lacking except in the young plants, whereas it becomes quite abundant when the Lettuce begins to "cabbage" and when it runs to seed. Now it is especially at the commencement of its life, at the budding of its first, tender leaves, that the plant needs to defend itself. One is inclined to think that the cultivated Lettuce loses its head a little, so to speak, and that it no longer knows exactly where it stands.

The Intelligence of the Flowers

of nuptial ceremonies that prevail in our gardens let us mention the ingenious ideas of some very simple flowers, in which the grooms and brides are born, love and die in the same corolla. The typical system is well enough known : the stamens, or male organs, generally frail and numerous, are grouped around the robust and patient pistil. But the disposition, the form, the habits of these organs vary in every flower, as though nature had a thought that cannot yet become settled, or an imagination that makes it a point of honour never to repeat itself. Often the pollen, when ripe, falls quite naturally from the top of the stamens upon the pistil; but very often, also, pistil and stamens are of the same height, or the latter are too far away, or the pistil is twice as tall as they. Then come endless efforts to succeed in meeting. Sometimes, as in the Nettle, the stamens, at the bottom of the corolla, stand cowering on their stalk: at the moment of fertilization, the stalk straightens out like a spring and the anther, or pollen-mass, that tops it shoots a cloud of dust over the stigma. Sometimes, as in the Barberry, whose nuptials can be accomplished only in the bright hours of a cloudless day, the stamens, far removed from the pistil, are kept against the sides of the flower by the weight of their moist glands: the sun appears and evaporates the fluid and the unballasted stamens dart upon the stigma. Elsewhere are different things again : thus, in the Primroses, the females are by turns longer and shorter than the males. In the Lily, the Tulip and other flowers, the too lanky bride does what she can to gather and fix the pollen. But the most original and fantastic system is that of the Rue (*Ruta graveolens*), a rather evil-smelling medicinal herb of the ill-famed emmenagogic tribe. The peaceful and docile stamens, drawn up in a circle around the fat, squat pistil, wait expectant in the yellow corolla. At the conjugal hour, obeying the command of the female, which apparently gives a sort of call by name, one of the males approaches and touches the stigma. Then come the third, the fifth, the seventh, the ninth male, until the whole row of odd numbers has rendered service. Next, in the even ranks, comes the turn of the second, the fourth, the sixth and so on. Here in verity is love to order! This flower which knows how to count appears to me so extraordinary that I at first refused to believe the botanists; and I was determined more than once to test its numerical sense before accepting it. I have ascertained positively that it but seldom makes a mistake.

It is superfluous to multiply these instances. A stroll in the woods or fields will allow any one to make a thousand observations in this direction, each quite as curious as those related by the botanists. But, before closing this chapter, I would mention one more flower: not that it displays any extraordinary imagination, but because of the delightful and easily-perceptible grace of its movement of love. I allude to the *Nigella Damascena*, or Fennel-flower, whose folk-names are charming : Love-in-a-mist, Devil-in-a-bush, Ragged-lady ; so many happy and touching efforts of popular poetry to describe a little flower that pleases it. This plant is found in a wild state in the South, by the roadside and under the olive-trees, and is

often cultivated in the North in old-fashioned
gardens. Its blossom is a pale blue, simple
as a floweret in a primitive painting, and the
"Venus' locks" or "ragged locks" that give
the Ragged-lady its popular name in France
are the light, tenuous, tangled leaves that
surround the corolla with a "bush" of misty
verdure. At the source of the flower, the
five extremely long pistils stand close-
grouped in the centre of the azure crown,
like five queens clad in green gowns,
haughty and inaccessible. Around them
crowd hopelessly the innumerous throng of
their lovers, the stamens, which do not come
up to their knees. And now, in the heart
of this palace of sapphires and turquoises, in
the gladness of the summer days, begins the
drama without words or catastrophe which
one might expect, the drama of powerless,
useless, motionless waiting. But the hours
pass that are the flower's years : its brilliancy
fades, its petals fall and the pride of the
great queens seems at last to bend under
the weight of life. At a given moment,
as though obeying the secret and irresistible
command of love, which deems the proof to

have lasted long enough, with a concerted
and symmetrical movement, comparable with
the harmonious parabolas of a five-fold jet
of water, they all together bend backwards,
stoop and gracefully cull the golden dust of
the nuptial kiss on the lips of their humble
lovers.

II

The unexpected abounds here, as we see.
A great volume might be written on the
intelligence of the plants, even as Romanes
wrote one on animal intelligence. But this
sketch has no pretension to become a
manual of that kind ; and I wish only to
call attention to a few interesting events
that happen beside us in this world wherein
we think ourselves, a little too vaiglori-
ously, privileged. These events are not
selected, but taken, by way of instances, as
the random result of observation and circumstance. I propose, however, in these
short notes, to concern myself above all with
the flower, for it is in the flower that the
greatest marvels shine forth. I set aside,
for the moment, the carnivorous flowers,

Life and Flowers

Droseras, Nepenthes and the rest, which approach the animal kingdom and would demand a special and expansive study, in order to devote myself to the true flower, the flower proper, which is believed to be motionless, insentient, passive and inanimate.

To separate facts from theories, let us speak of the flower as though all that it has realized had been foreseen and conceived in the manner of men. We shall see later how much we must leave to it, how much take away from it. For the present, let it take the stage alone, like a splendid princess endowed with reason and will. There is no denying that it appears to be provided with both; and to deprive it of either we should have to resort to very obscure hypotheses. It is there, then, motionless on its stalk, sheltering in a dazzling tabernacle the reproductive organs of the plant. Apparently, it has but to allow the mysterious union of the stamens and pistil to be accomplished in this tabernacle of love. And many flowers do so consent. But to many others there is propounded, big with awful threats, the normally insoluble problem of cross-fertilization. As the result of what numberless and immemorial experiments did they observe that self-fertilization —that is the fertilization of the stigma by the pollen falling from the anthers that surround it in the same corolla—rapidly induces the degeneration of the species? They have observed nothing, we are told, nor profited by any experience. The force of things quite simply and gradually eliminated the seeds and plants weakened by self-fertilization. Soon only those survived which, through some anomaly, such as the exaggerated length of the pistil, rendering it inaccessible to the anthers, were prevented from fertilizing themselves. These exceptions alone endured, through a thousand revolutions; heredity finally determined the work of chance; and the normal type disappeared.

12

We shall see presently what light these explanations throw. For the moment, let us

stroll into the garden or the field, to study
more closely two or three curious inventions
of the genius of the flower. And already,
without going far from the house, we have
here, frequented by the bees, a sweet-
scented cluster inhabited by a most skilled
mechanic. There is no one, even among
the least countrified, but knows the good
Sage. It is an unpretending *Labiata* and
bears a very modest flower, which opens
violently, like a hungry mouth, to snap the
rays of the sun in passing. For that matter,
it presents a large number of varieties,
not all of which—this is a curious detail—
have adopted or carried to the same pitch
of perfection the system of fertilization
which we are about to examine. But I am
concerned here only with the most common
Sage, that which, at this moment, as though
to celebrate spring's passage, covers with
violet draperies all the walls of my terraces
of olive-trees. I assure you that the bal-
conies of the great marble palaces that await
the kings were never more luxuriously, more
happily, more fragrantly adorned. One
seems to catch the very perfumes of the

light of the sun at its hottest, when noon-day strikes. . . .

To come to details, the stigma, or female organ, of the flower is contained in the upper lip, which forms a sort of hood, in which are also the two stamens, or male organs. To prevent these from fertilizing the stigma which shares the same nuptial tent, this stigma is twice as long as they, so that they have no hope of reaching it. Moreover, in order to avoid any accident, the flower has made itself protenandrous, that is to say, the stamens ripen before the pistil, so that, when the female is fit to conceive, the males have already disappeared. It is necessary, therefore, that an external force should intervene to accomplish the union by carrying a foreign pollen to the abandoned stigma. A certain number of flowers, the anemophilous flowers, leave this care to the wind. But the Sage—and this is the more general case—is entomophilous, that is to say, it loves insects and relies upon their collaboration alone. Still, it is quite aware, for it knows many things, that it lives in a world where it is best to expect no sympathy, no charitable aid. It does not waste time, therefore, in making useless appeals to the courtesy of the bee. The bee, like all that struggles against death in this world of ours, exists only for herself and for her kind and is in no way concerned to render a service to the flowers that feed her. How shall she be obliged, in spite of herself, or at least unconsciously, to fulfil her matrimonial office? Observe the wonderful love-trap contrived by the Sage: right at the back of its tent of violet silk, it distils a few drops of nectar; this is the bait. But, barring the access to the sugary fluid, stand two parallel stalks, somewhat similar to the uprights of a Dutch drawbridge. Right at the top of each stalk is a great sack, the anther, overflowing with pollen; at the bottom, two smaller sacks serve as a counterpoise. When the bee enters the flower, in order to reach the nectar she has to push the small sacks with her head. The two stalks, which turn on an axis, at once topple over and the upper anthers come down and touch the sides of the insect, whom they

The Intelligence of the Flowers

cover with fertilizing dust. No sooner has the bee departed than the springy pivots fly back and replace the mechanism in its first position; and all is ready to repeat the work at the next visit.

However, this is only the first half of the play: the sequel is enacted in another scene. In a neighbouring flower, whose stamens have just withered, enters upon the stage the pistil that awaits the pollen. It issues slowly from the hood, lengthens out, stoops, curves down, becomes forked so as, in its turn, to bar the entrance to the tent. On its way to the nectar, the head of the bee passes freely under the hanging fork, which, however, grazes her back and sides exactly at the spots touched by the stamens. The two-cleft stigma greedily absorbs the silvery dust; and the impregnation is accomplished. It is easy, for that matter, by introducing a straw or the end of a match, to set the apparatus going and to take stock of the striking and marvellous combination and precision of all its movements.

The varieties of the Sage are very many

—they number about five hundred—and I will spare you the majority of their scientific names, which are not always pretty: *Salvia pratensis*, *officinalis* (our Garden Sage), *Horminum*, *Horminoides*, *glutinosa*, *Sclarea*, *Roemeri*, *azurea*, *Pitcheri*, *splendens* (the magnificent Sage of our baskets) and so on. There is not, perhaps, one but has modified some detail of the machinery which we have just examined. A few—and this, I think, is a doubtful improvement—have doubled and sometimes trebled the length of the pistil, so that it not only emerges from the hood, but makes a wide plume-like curve in front of the entrance to the flower. They thus avoid the just-possible danger of the fertilization of the stigma by the anthers dwelling in the same hood; but, on the other hand, it may happen, if the protenandry be not strict, that the insect, on leaving the flower, deposits on the stigma the pollen of the very anthers with which the stigma cohabits. Others, in the movement of the lever, make the anthers diverge farther apart so as to strike the sides of the animal with greater precision. Others, lastly, have not

244

The Intelligence of the Flowers

succeeded in arranging and adjusting every part of the mechanism. I find, for instance, not far from my violet Sage, near the well, under a cluster of Oleanders, a family of white flowers tinted with pale lilac which have no suggestion or trace of a lever. The stamens and the stigma are heaped up promiscuously in the middle of the corolla. All seems left to chance and disorganized.

I have no doubt that it would be possible, to any one collecting the very numerous varieties of this *Labiata*, to reconstruct the whole history, to follow all the stages of the invention, from the primitive disorder of the white Sage under my eyes to the latest improvements of the *Salvia pratensis*. What conclusion are we to draw? Is the system still in the experimental stage among the aromatic tribe? Has it not yet left the period of models and "trial trips," as in the case of the Archimedean screw in the Saintfoin family? Has the excellence of the automatic lever not yet been unanimously admitted? Can it be, then, that everything is not unchangeable and preestablished ; and are they still discussing and

experimenting in this world which we believe to be fatally, organically regular ?[1]

13

Be this as it may, the flower of most
varieties of the Sage presents an attractive
solution of the great problem of cross-
fertilization. But, even as, among men, a
new invention is at once taken up, simplified, improved by a host of small indefati-

---

[1] For nearly four years, I have been engaged upon
a series of experiments in the hybridization of Sages,
artificially fertilizing (first taking the usual precautions
against any interference of wind or insects) a variety of
which the floral mechanism has reached a high state of
perfection with the pollen of a very backward variety;
and *vice versa*. My observations are not yet sufficiently
numerous to permit me to give any details or conclusions
here. Nevertheless, it already appears as if a general
law were being evolved, namely that the backward Sage
readily adopts the improvements of the more advanced
variety, whereas the latter is not so prone to accept the
defects of the first. This would tend to throw an interesting side-light upon the operations, the habits, the
preferences, the tastes of nature at her best. But these
experiments cannot possibly be completed in so short a
period, because of the time lost in collecting the different
varieties, of the numberless proofs and counter-proofs
required and so on. It would be premature, therefore,
as yet to draw the slightest conclusion from them.

gable seekers, so, in the world of what we
may call mechanical flowers, the patent of
the Sage has been elaborated and in many
details strangely perfected. A pretty general
Scrophularinea, the common Lousewort, or
Red-rattle (*Pedicularis sylvatica*), which you
must surely have noticed in the shady parts
of small woods and heaths, has introduced
some extremely ingenious modifications.
The shape of the corolla is almost similar
to that of the Sage ; the stigma and the two
anthers are all three contained in the upper
hood. Only the little moist tip of the pistil
protrudes from the hood, while the anthers
remain captive.   In this silky tabernacle,
therefore, the organs of the two sexes are
very close together and even in immediate
contact ; nevertheless, thanks to an enact-
ment quite different from that of the Sage,
self-fertilization is made absolutely impossible.
The anthers, in fact, form two sacks filled
with powder ; each of the sacks has only one
opening and they are juxtaposed in such a
way that the openings coincide and mutually
close each other.   They are forcibly kept
inside the hood, on their curved, springy

stalks, by a sort of teeth. The bee or
humble-bee that enters the flower to sip its
nectar necessarily pushes these teeth aside;
and the sacks are no sooner set free than
they fly up, are flung outside and alight
upon the back of the insect.

But the genius and foresight of the flower
go farther than this. As Hermann Müller,
who was the first to make a complete study
of the wonderful mechanism of the Lousewort, observes (I am quoting from a summary):

"If the stamens struck the insect while
preserving their relative positions, not a
grain of pollen would leave them, because
their orifices reciprocally close each other.
But a contrivance which is as simple as it
is ingenious overcomes the difficulty. The
lower lip of the corolla, instead of being
symmetrical and horizontal, is irregular and
slanting, so that one side of it is higher
by a few millimetres than the other. The
humble - bee resting upon it must herself
necessarily stand in a sloping position. The
result is that her head strikes first one and

then the other of the projections of the corolla. Therefore the releasing of the stamens also takes place successively; and, one after the other, their orifices, now freed, strike the insect and sprinkle her with fertilizing dust.

"When the humble-bee next passes to another flower, she inevitably fertilizes it, because—and I have purposely omitted this detail—what she meets first of all, when thrusting her head into the entrance to the corolla, is the stigma, which grazes her just at the spot where she is about, the moment after, to be struck by the stamens, the exact spot where she has already been touched by the stamens of the flower which she has last left."

14

These instances might be multiplied indefinitely; every flower has its idea, its system, its acquired experience which it turns to advantage. When we examine closely their little inventions, their diverse methods, we are reminded of those enthralling exhibitions of machine-tools, of machines for

making machinery, in which the mechanical
genius of man reveals all its resources. But
our mechanical genius dates from yesterday,
whereas floral mechanism has been at work
for thousands of years. When the flowers
made their appearance upon our earth, there
were no models around them which they
could imitate; they had to derive everything from within themselves. At the
period when we had not gone beyond the
club, the bow and the flail; in the com-
paratively recent days when we conceived
the spinning-wheel, the pulley, the tackle,
the ram; at the time—it was last year, so
to speak—when our master-pieces were the
catapult, the clock and the weaving-loom,
the Sage had contrived the uprights and
counterweights of its lever of precision and
the Lousewort its sacks closed up as though
for a scientific experiment, the successive
releasing of its springs and the combination
of its inclined planes. Who, say a hundred
years ago, dreamt of the properties of the
screw which the Maple and the Lime-tree
have been turning to use since the birth of the
trees? When shall we succeed in building

The Intelligence of the Flowers

a parachute or a flying-machine as rigid, as
light, as subtle and as safe as that of the
Dandelion? When shall we discover the
secret of cutting in so frail a fabric as the
silk of the petals a spring as powerful as
that which projects into space the golden
pollen of the Spanish Broom? As for the
Momordica, or Squirting Cucumber, whose
name I mentioned at the beginning of this
little study, who shall tell us the mystery
of its miraculous, strength? Do you know
the Momordica? It is a humble *Cucurbi-*
tacea, common enough along the Medi-
terranean coast. Its prickly fruit, which
resembles a small cucumber, is endowed with
inexplicable vitality and energy. You have
but to touch it, at the moment of its ma-
turity, and it suddenly quits its peduncle by
means of a convulsive contraction and shoots
through the hole produced by the wrench,
mingled with numerous seeds, a mucila-
ginous stream of such wonderful intensity
that it carries the seed to four or five yards'
distance from the natal plant. The action
is as extraordinary, in proportion, as though
we were to succeed in emptying ourselves

with a single spasmodic movement and in precipitating all our organs, our viscera and our blood to a distance of half a mile from our skin and skeleton.

A large number of seeds besides have ballistic methods and employ sources of energy that are more or less unknown to us. Remember, for instance, the explosions of the Colza and the Heath. But one of the great masters of vegetable artillery is the Spurge. The Spurge is an *Euphorbiacea* of our climes, a tall and fairly ornamental "weed," which often exceeds the height of a man. I have a branch of Spurge on my table at this moment steeped in a glass of water. It has trifid, greenish berries, which contain the seeds. From time to time, one of these berries bursts with a loud report; and the seeds, gifted with a prodigious initial velocity, strike the furniture and the walls on every side. If one of them hits your face, you feel as though you had been stung by an insect, so extraordinary is the penetrating force of these tiny seeds, each no larger than a pin's head. Examine the berry, look for the springs that give it life : you shall not find the secret of this force, which is as invisible as that of our nerves.

The Spanish Broom (*Spartium junceum*) has not only pods, but flowers fitted with springs. You may have remarked the wonderful plant. It is the proudest representative of this powerful family of the Brooms. Greedy of life, poor, sober, robust, rejecting no soil, no trial, it forms along the paths and in the mountains of the South huge, tufted balls, sometimes three yards high, which, between May and June, are covered with a magnificent bloom of pure gold, whose perfumes, mingling with those of its habitual neighbour, the Honeysuckle, spread under the fury of a fierce sun delights that are not to be described save by evoking celestial dews, Elysian springs, cool streams and starry transparencies in the hollow of azure grottoes. . . .

The flower of this Broom, like that of all the papilionaceous *Leguminosæ*, resembles the flowers of the Peas of our gardens ; and its lower petals, shaped like the beak of a galley, contain hermetically the stamens and the pistil. So long as it is not ripe, the bee who explores it finds it impenetrable. But, as soon as the moment of puberty arrives for the captive bride and grooms, the beak bends under the weight of the insect that rests upon it; and the golden chamber bursts voluptuously, hurling with violence and afar, over the visitor, over the flowers around, a cloud of luminous dust, which a broad petal, shaped like a penthouse, casts down upon the stigma to be impregnated.

15

Let us leave the seeds and return to the flowers. As I have said, one could prolong indefinitely the list of their ingenious inventions. I refer those who might wish to study these problems thoroughly to the works of Christian Konrad Sprengel, who was the first, in 1793, in his curious volume, *Das entdeckte Geheimniss der Natur im Bau und in der Befruchtung der Blumen*, to analyze the functions of the different organs

The Intelligence of the Flowers

in the Orchids; next, to the books of
Charles Darwin, Dr. Hermann Müller of
Lippstadt, Hildebrand, Delpino the Italian,
Sir William Hooker, Robert Brown and
many others.

We shall find the most perfect and the
most harmonious manifestations of vegetable intelligence among the Orchids. In
these writhing and eccentric flowers, the
genius of the plant touches its extreme
point and with an unusual fire pierces
the wall that separates the kingdoms. For
that matter, this name of Orchid must
not be allowed to mislead us or make us
believe that we have here to do only with
rare and precious flowers, with those hot-
house queens which seem to claim the care
of the goldsmith rather than the gardener.
Our native wild flora, which comprises all
our modest "weeds," numbers more than
twenty-five species of Orchids, including
just the most ingenious and complicated.
It is these which Charles Darwin studied
in his book, *On the Various Contrivances by
which Orchids are fertilized by Insects*, which
is the wonderful history of the most heroic

efforts of the soul of the flower. It is out of the question that I should here, in a few lines, summarize that abundant and fairylike biography. Nevertheless, since we are on the subject of the intelligence of flowers, it is necessary that we should give some idea of the methods and the mental habits of that which excels all the others in the art of compelling the bee or the butterfly to do exactly what it wishes, in the prescribed form and time.

16

It is not easy to explain without diagrams the extraordinarily complex mechanism of the Orchid. Nevertheless, I will try to give a sufficient idea of it with the aid of more or less approximate comparisons, while avoiding as far as possible the use of technical terms such as *retinaculum*, *labellum*, *rostellum* and the rest, which evoke no precise image in the minds of persons unfamiliar with botany.

Let us take one of the most widely-distributed Orchids in our regions, the

The Intelligence of the Flowers

Orchis maculata, for instance, or rather, because it is a little larger and therefore more easily observed, the *Orchis latifolia*, the Marsh Orchid, commonly known as the Meadow-rocket. It is a perennial plant and grows to a height of an inch or more. It is fairly common in the woods and damp meadows and it bears a thyrse of little pink flowers which blossom in May and June. The typical flower of our Orchids represents pretty closely the fantastic and yawning mouth of a Chinese dragon. The lower lip, which is very long and which hangs in the form of a jagged or dentate apron, serves as a landing-place for the insect. The upper lip rounds into a sort of hood, which shelters the essential organs; while, at the back of the flower, beside the peduncle, there falls a kind of spur or long, pointed horn, which contains the nectar. In most flowers, the stigma, or female organ, is a more or less viscid little tuft which, at the end of a frail stalk, patiently awaits the coming of the pollen. In the Orchid, this traditional installation has become irrecognizable. At the back of the mouth, in the place occupied in the throat by the uvula, are two closely-welded stigmas, above which rises a third stigma modified into an extraordinary organ. At its top, it carries a sort of little pouch, or, more correctly, a sort of stoup, which is called the *rostellum*. This stoup is full of a viscid fluid in which soak two tiny balls whence issue two short stalks laden at their upper extremity with a packet of grains of pollen carefully tied up.

Let us now see what happens when an insect enters the flower. She lands on the lower lip, outspread to receive her, and, attracted by the scent of the nectar, seeks to reach the horn that contains it, right at the back. But the passage is purposely very narrow; and the insect's head, as she advances, necessarily strikes the stoup. The latter, sensitive to the least shock, is at once ruptured along a convenient line and lays bare the two little balls steeped in the viscid fluid. These, coming into immediate contact with the visitor's skull, fasten to it and become firmly stuck to it, so that, when the insect leaves the flower, she carries them away and, with them, the two

The Intelligence of the Flowers

stalks which rise from them and which
end in the packets of tied-up pollen. We
therefore have the insect capped with two
straight, bottle-shaped horns. The unconscious artisan of a difficult work now visits
a neighbouring flower. If her horns remained stiff, they would simply strike with
their pollen-masses the other pollen-masses
soaking in the vigilant stoup and no event
would spring from the pollen mingling with
pollen. But here the genius, the experience
and the foresight of the Orchid become apparent. The Orchid has minutely calculated
the time needed for the insect to suck the
nectar and repair to the next flower; and it
has ascertained that this requires, on an average, thirty seconds. We have seen that the
packets of pollen are carried on two short
stalks inserted into the viscid balls. Now
at the point of insertion there is, under each
stalk, a small membranous disc, whose only
function is, at the end of thirty seconds,
to contract and throw forward the stalks,
causing them to curve and describe an
arc of ninety degrees. This is the result
of a fresh calculation, not of time, on this

occasion, but of space. The two horns of
pollen that cap the nuptial messenger are
now horizontal and point in front of her
head, so that, when she enters the next
flower, they will just strike the two welded
stigmas under the pendent stoup.

This is not all and the genius of the
Orchid has not yet expended all its foresight. The stigma which receives the blow
of the packet is coated with a viscid substance. If this substance were as powerfully
adhesive as that contained in the stoup, the
pollen-masses, after their stalks were broken,
would stick to it and remain fixed to it
whole; and their destiny would be ended.
This must not be; it is important that the
chances of the pollen should not be exhausted
in a single venture, but rather that they
should be multiplied to the greatest possible
extent. The flower that counts the seconds
and measures the lines is a chemist to boot
and distils two sorts of gums: one extremely
clinging, hardening as soon as it touches
the air, to glue the pollen-horns to the
insect's head; the other greatly lenified, for
the work of the stigma. This latter is just

The Intelligence of the Flowers

prehensile enough slightly to unfasten or loosen the tenuous and elastic threads with which the grains of pollen are tied up. Some of these grains cling to it, but the pollinic mass is not destroyed; and, when the insect visits other flowers, she will continue her fertilizing labours almost indefinitely.

Have I expounded the whole miracle? No; I have still to call attention to many a neglected detail: among others, to the movement of the little stoup, which, after its membrane has been ruptured to unmask the viscid balls, immediately lifts up its lower rim in order to keep in good condition, in the sticky fluid, the packet of pollen which the insect may not have carried off. We should also note the very curiously-combined divergence of the pollinic stalks on the head of the insect, as well as certain chemical precautions common to all plants; for the experiments made quite recently by M. Gaston Bonnier seem to prove that every flower, in order to preserve its species intact, secretes poisons that destroy or sterilize any foreign pollen. This

is about all that we see ; but here, as in all
things, the real, the great miracle begins
where our power of vision ends.

17

I have just this moment found, in an
untilled corner of the olive-yard, a splendid
sprig of *Loroglossum hircinum*, a variety
which, for I know not what reason (perhaps it is very rare in England), Darwin
omitted to study. It is certainly the most
remarkable, the most fantastic, the most
astounding of all our native Orchids. If it
were of the size of the American Orchids,
one could declare that there is no more
fanciful plant in existence. Imagine a thyrse,
like that of the Hyacinth, but twice as tall.
It is symmetrically adorned with ill-favoured,
three-cornered flowers, of a greenish white
stippled with pale violet. The lower petal,
embellished at its source with bronzed caruncles, huge mustachios and sinister-looking
lilac buboes, stretches out interminably, madly,
unreally, in the shape of a corkscrew riband
of the colour assumed by a drowned corpse

after a month's immersion in the river. From the whole, which conjures up the idea of the most fearsome maladies and seems to blossom in some vague land of ironical nightmares and witcheries, there issues a potent and abominable stench as of a poisoned goat, which spreads afar and reveals the presence of the monster. I am pointing to and describing this nauseating Orchid because it is fairly common in France, is easily recognized and adapts itself very well, by reason of its height and the distinctness of its organs, to any experiments that one might wish to make. We have only, in fact, to introduce the tip of a match into the flower and to push it carefully to the bottom of the nectary, in order, with the naked eye, to witness all the successive revolutions of the process of fertilization. Grazed in passing, the pouch or *rostellum* sinks down, exposing the little viscid disc (the *Loroglossum* has only one) that supports the two pollen-stalks. As soon as this disc violently grips the end of the wood, the two cells that contain the pollen-balls open longitudinally; and, when the match is withdrawn, its tip is firmly capped with two stiff, diverging horns, ending in two golden balls. Unfortunately, we do not here, as in the experiment with the *Orchis latifolia*, enjoy the charming spectacle offered by the gradual and precise inclination of the two horns. Why are they not lowered? We have but to introduce the capped match into a neighbouring nectary to ascertain that this movement would be superfluous, the flower being much larger than that of the *Orchis maculata* or *latifolia* and the nectar-horn arranged in such a way that, when the insect laden with the pollen-masses enters it, they just reach the level of the stigma to be fertilized.

Let us add that it is important to the success of the experiment to select a flower that is quite ripe. We do not know when the flower is ripe; but the insect and the flower itself know, for the flower does not invite its necessary guests, by offering them a drop of nectar, until the moment comes when all its apparatus is ready to work.

The Intelligence of the Flowers

18

This is the basis of the system of fertilization adopted by the Orchid of our climes. But each species, every family modifies and improves the details in accordance with its particular experience, psychology and convenience. The *Orchis* or *Anacamptis pyramidalis*, for instance, which is one of the most intelligent, has added to its lower lip or *labellum* two little ridges which guide the proboscis of the insect to the nectar and compel her to accomplish exactly what is expected of her. Darwin very justly compares this ingenious accessory with the little instrument for guiding a thread into the fine eye of a needle. Here is another interesting improvement: the two little balls that carry the pollen-stalks and soak in the stoup are replaced by a single viscid disc, shaped like a saddle. If, following the road to be taken by the insect's proboscis, we insert the point of a needle or a bristle into the flower, we very plainly perceive the advantages of this simpler and more practical arrangement. As the bristle touches

the stoup, the latter is ruptured in a symmetrical line and uncovers the saddle-formed
disc, which at once becomes attached to the
bristle. Withdraw the bristle smartly and
you will just have time to catch the pretty
action of the saddle, which, seated on the
bristle or needle, curls its two flaps inwards,
so as to embrace the object that supports
it. The purpose of this movement is to
strengthen the adhesive power of the saddle
and, above all, to ensure with greater precision than in the *Orchis latifolia* the indispensable divergence of the pollen-stalks.
As soon as the saddle has curled round the
bristle and as the pollen-stalks planted in it,
drawn apart by its contraction, are forced to
diverge, the second movement of the stalks
begins and they bend towards the tip of the
bristle, in the same manner as in the Orchid
which we have already studied. The two
movements combined are performed in thirty
to thirty-four seconds.

19

Is it not exactly in this manner, by means
of trifles, of successive overhaulings and

The Intelligence of the Flowers

retouches, that human inventions proceed? We have all, in the latest of our mechanical industries, followed the tiny, but constant improvements in the sparking, the carburation, the clutch and the speed-gear. It would really seem as though ideas came to the flowers in the same way as to us. The flowers grope in the same darkness, encounter the same obstacles, the same ill-will, in the same unknown. They have the same laws, the same disillusions, the same slow and difficult triumphs. They would appear to possess our patience, our perseverance, our self-love, the same varied and diversified intelligence, almost the same hopes and the same ideals. They struggle, like ourselves, against a great indifferent force that ends by assisting them. Their inventive imagination not only follows the same prudent and minute methods, the same tiring, narrow and winding little paths: it also has unexpected leaps and bounds that suddenly fix definitely an uncertain discovery. It is thus that a family of great inventors among the Orchids, a strange and rich American family, that of the *Catasetidæ*, thanks to a bold

267

inspiration, abruptly altered a number of habits
that doubtless appeared to it too primitive.
First of all, the separation of the sexes is absolute : each has its particular flower. Next,
the *pollinium*, or mass or packet of pollen,
no longer dips its stalk in a stoup full of
gum, there awaiting, a little inertly and, in
any case, without initiative, the lucky accident that is to fix it on the insect's head.
It is bent back on a powerful spring, in a
sort of cell. Nothing attracts the insect
specially in the direction of this cell. Nor
have the proud *Catasetidæ* reckoned, like the
common Orchid, on this or that movement
of the visitor : a guided and precise movement, if you wish, but nevertheless a con-
tingent movement. No, the insect no longer
enters a flower endowed with an admirable
mechanism : she enters an animated and literally sensitive flower. Hardly has she pitched
upon the magnificent outer court of copper-
coloured silk before long and nervous
feelers, which she cannot avoid touching,
carry the alarm all over the edifice. Forthwith the cell is torn asunder in which the
pollen-mass, divided into two packets, is

held captive on its bent-back pedicel, which is supported on a huge viscid disc. Abruptly released, the pedicel straightens itself like a spring, dragging with it the two packets of pollen and the viscid disc, which are violently projected outside. In consequence of a curious ballistic calculation, the disc is always flung first and strikes the insect, to whom it adheres. She, stunned by the blow, has but one thought: to leave the aggressive corolla with all speed and take refuge in a neighbouring flower. This is all that the American Orchid wanted.

20

Shall I describe also the curious and practical simplifications introduced into the general system by another family of exotic Orchids, the *Cypripedeæ?* Let us continue to bear in mind the circumvolutions of human inventions: we have here an amusing counter-proof. A fitter, in the engine-room, a preparator, a pupil, in the laboratory, says, one day, to his principal:

"Suppose we tried to do just the opposite?

Life and Flowers

Suppose we reversed the movement, suppose we inverted the mixture of the fluids?"

The experiment is tried; and suddenly from the unknown issues the unexpected.

One could easily believe the *Cypripedeæ* to have held similar conversations among themselves. We all know the Cypripedium, or Ladies'-slipper: with its enormous shoe-shaped chin, its crabbed and venomous air, it is the most characteristic flower of our hothouses, the one that seems to us the typical Orchid, so to speak. The Cypripedium has bravely suppressed all the complicated and delicate apparatus of the springy pollen-masses, the diverging stalks, the viscid discs, the cunning gums and the rest. Its clog-like chin and a barren, shield-shaped anther bar the entrance in such a manner as to compel the insect to pass its proboscis over two little heaps of pollen. But this is not the important point: the wholly unexpected and abnormal thing is that, contrary to what we have observed in all the other species, it is no longer the stigma, the female organ that is viscid, but the pollen itself, whose grains, instead of being pulverulent, are covered with a coat so glutinous that it can be stretched and drawn into threads. What are the advantages and the drawbacks of this new arrangement? It is to be feared that the pollen carried off by the insect may adhere to any object other than the stigma; on the other hand, the stigma is dispensed from secreting the fluid destined to sterilize every foreign pollen. In any case, this problem would demand a special study. In the same way, there are patents whose usefulness we do not grasp at once.

21

To have done with this strange tribe of the Orchids, it remains for us to say a few words on an auxiliary organ that sets the whole mechanism going: I mean the nectary, which, for that matter, has been the object, on the part of the genius of the species, of enquiries, attempts and experiments as intelligent and as varied as those which are incessantly modifying the economy of the essential organs.

Life and Flowers

The nectary, as we have seen, is, in principle, a sort of spur, of long, pointed horn that opens right at the bottom of the flower, beside the peduncle, and acts more or less as a counterpoise to the corolla. It contains a sugary liquid, the nectar, which serves as food for butterflies, beetles and other insects and which is turned into honey by the bee. Its business, therefore, is to attract the indispensable guests. It is adapted to their size, their habits, their tastes; it is always arranged in such a way that they cannot introduce or withdraw their proboscis without scrupulously and successively performing all the rights prescribed by the organic laws of the flower.

We already know enough of the fantastic character and imagination of the Orchids to foresee that here, as elsewhere—and even more than elsewhere, for the more supple organ lends itself to this more readily—their inventive, practical, observant and groping spirit has given itself free scope. One of them, for instance, the *Sarcanthus teretifolius*, probably failing in its endeavour to elaborate a viscid fluid that should harden quickly enough

The Intelligence of the Flowers

to stick the bundle of pollen to the insect's head, has overcome the difficulty by delaying the visitor's proboscis as long as possible in the narrow passages leading to the nectar. The labyrinth which it has laid out is so complicated that Bauer, Darwin's skilful draughtsman, had to admit himself beaten and gave up the attempt to reproduce it.

There are some which, starting on the excellent principle that every simplification is an improvement, have boldly suppressed the nectar-horn. They have replaced it by certain fleshy, fantastic and evidently succulent excrescences which are nibbled by the insects. Is it necessary to add that these excrescences are always placed in such a manner that the guest who feasts on them must inevitably set all the pollen-machinery in movement?

22

But, without lingering over a thousand very various little artifices, let us end these fairy stories by studying the enticements of the *Coryanthes macrantha*. Truly, we

no longer know with exactly what sort of
being we here have to do. The astounding
Orchid has contrived this: its lower lip or
labellum forms a sort of bucket, into which
drops of almost pure water, secreted by two
horns situated overhead, fall continually;
when this bucket is half full, the water
flows away on one side by a spout or gutter.
All this hydraulic installation is very remarkable in itself; but here is where the
alarming, I might almost say the diabolical
side of the combination begins. The liquid
which is secreted by the horns and which
accumulates in the satin basin is not nectar
and is in no way intended to attract the
insects: it has a much more delicate function in the really machiavellian plan of this
strange flower. The artless insects are in-
vited by the sugary perfumes diffused by
the fleshy excrescences of which I spoke
above to walk into the trap. These excrescences are above the cup, in a sort of
chamber to which two lateral openings give
access. The big visiting bee—the flower,
being enormous, allures hardly any but the
heaviest *Hymenopteræ*, as though the others

experienced a certain shame at entering such vast and sumptuous halls—the big bee begins to nibble the savoury caruncles. If she were alone, she would go away quietly, after finishing her meal, without even grazing the bucket of water, the stigma and the pollen; and none of that which is required would take place. But the sapient Orchid observes the life that moves around it. It knows that the bees form an innumerable, greedy and busy people, that they come out by thousands in the sunny hours, that a perfume has but to quiver like a kiss on the threshold of an opening flower for them to hasten in a crowd to the banquet prepared under the nuptial tent. We therefore have two or three looters in the sugary chamber: the space is scanty, the walls slippery, the guests ill-mannered. They crowd and hustle one another to such good purpose that one of them always ends by falling into the bucket that awaits her beneath the treacherous repast. She there finds an unexpected bath, conscientiously wets her bright, diaphanous wings and, despite immense efforts, cannot

succeed in resuming her flight. This is
where the astute flower lies in wait for her.
There is but one opening through which she
can leave the magic bucket: the spout that
acts as a waste-pipe for the overflow of the
reservoir. It is just wide enough to allow
of the passage of the insect, whose back
touches first the sticky surface of the stigma
and then the viscid glands of the pollen-
masses that await her along the vault. She
thus escapes, laden with the adhesive dust,
and enters a neighbouring flower, where the
tragedy of the banquet, the hustling, the
fall, the bath and the escape is reenacted
and perforce brings the imported pollen
into contact with the greedy stigma.

Here, then, we have a flower that knows
and plays upon the passions of insects. Nor
can it be pretended that all these are only
so many more or less romantic interpretations: no, the facts have been precisely
and scientifically observed and it is impossible to explain the use and arrangement of the flower's different organs in
any other way. We must accept the evidence as it stands. This incredible and

The Intelligence of the Flowers

efficacious artifice is the more surprising inasmuch as it does not here tend to satisfy the immediate and urgent need to eat that sharpens the dullest wits; it has only a distant ideal in view: the propagation of the species.

But why, we shall be asked, these fantastic complications which end only by increasing the dangers of chance? Let us not hasten to give judgment and reply. We know nothing of the reasons of the plant. Do we know what obstacles the flower encounters in the direction of logic and simplicity? Do we know thoroughly a single one of the organic laws of its existence and its growth? One watching us from the height of Mars or Venus, as we exert ourselves to achieve the conquest of the air, might, in his turn, ask:

"Why those shapeless and monstrous machines, those balloons, those air-ships, those parachutes, when it were so easy to copy the birds and to supply the arms with a pair of all-sufficing wings?"

To these proofs of intelligence, man's somewhat puerile vanity opposes the traditional objection: yes, they create marvels, but those marvels remain eternally the same. Each species, each variety has its system and, from generation to generation, introduces no perceptible improvement. It is true that, since we have been observing them—that is to say, during the past fifty years—we have not seen the *Coryanthes macrantha* or the *Catasetidæ* perfect their trap: this is all that we can say; and it is really not enough. Have we as much as attempted the most elementary experiments; and do we know what the successive generations of our astonishing bathing Orchid might do in a century's time, if placed in different surroundings, among insects to which it was not accustomed? Besides, the names which we give to the orders, species and varieties end by deceiving ourselves; and we thus create imaginary types which we believe to be fixed, whereas they are probably only the representatives of one

The Intelligence of the Flowers

and the same flower, which continues to modify its organs slowly in accordance with slow circumstances.

The flowers came upon our earth before the insects; they had, therefore, when the latter appeared, to adapt an entirely new system of machinery to the habits of these unexpected collaborators. This geologically-incontestable fact alone, amid all that which we do not know, is enough to establish evolution; and does not this somewhat vague word mean, after all, adaptation, modification, intelligent progress?

It would be easy, moreover, without appealing to this prehistoric event, to bring together a large number of facts which would show that the faculty of adaptation and intelligent progress is not reserved exclusively for the human race. Without returning to the detailed chapters which I have devoted to this subject in *The Life of the Bee*, I will simply recall two or three topical details which are there mentioned. The bees, for instance, invented the hive. In the wild and primitive state and in their country of origin, they work in the

Open air. It was the uncertainty, the inclemency of our northern seasons that gave them the idea of seeking a shelter in hollow trees or a hole in the rocks. This ingenious idea restored to the work of looting and to the care of the eggs the thousands of bees stationed around the combs to maintain the necessary heat. It is not uncommon, especially in the South, during exceptionally mild summers, to find them reverting to the tropical manners of their ancestors.[1]

Another fact: when transported to Australia or California, our black bee completely alters her habits. After one or two years, finding that summer is perpetual and flowers for ever abundant, she will live from

[1] I had just written these lines, when M. E. L. Bouvier made a communication in the Academy of Science (*cf.* the report of the 7th of May 1906) on the subject of two nidifications in the open air observed in Paris, one in a *Sophora Japonica*, the other in a chestnut-tree. The latter, which hung from a small branch furnished with two almost contiguous forks, was the more remarkable of the two, because of its evident and intelligent adaptation to particularly difficult circumstances.

"The bees," says M. de Parville, in his summary in the science-column of the *Journal des Débats* of the

day to day, content to gather the honey and pollen indispensable for the day's consumption; and, her recent and thoughtful observation triumphing over hereditary experience, she will cease to make provision for her winter. Büchner mentions an analogous fact, which also proves the bees' adaptation to circumstances, not slow, secular, unconscious and fatal, but immediate and intelligent: in Barbados, the bees whose hives are in the midst of the refineries, where they find sugar in plenty during the whole year, will entirely abandon their visits to the flowers.

Let us lastly recall the amusing contradiction which the bees gave to two learned English entomologists, Kirkby and Spence:

"Show us," said these, "a single case in

31st of May 1906, "built consolidating pillars and resorted to really remarkable artifices of protection and ended by transforming the two forks of the chestnut-tree into a solid ceiling. An ingenious human being would certainly not have done so well.

"To protect themselves against the rain, they had installed fences, thickenings and blinds against the sun. One can have no idea of the perfection of the industry of the bees, except by observing the architecture of the two nidifications, now at the Museum."

Life and Flowers

which, under stress of circumstances, the bees have had the idea of substituting clay or mortar for wax and propolis and we will admit their reasoning faculties."

Hardly had they expressed this somewhat arbitrary wish, when another naturalist, Andrew Knight, having coated the bark of certain trees with a sort of cement made of wax and turpentine, observed that his bees entirely ceased to gather propolis and employed only this new and unknown substance, which they found prepared in abundance in the neighbourhood of their home. Moreover, in the practice of apiculture, when pollen is scarce, the bee-keeper has but to place a few handfuls of flour at their disposal for them at once to understand that this can serve the same purpose and be turned to the same use as the dust of the anthers, although its taste, smell and colour are absolutely different.

What I have just said, in the matter of the bees, might, I think, *mutatis mutandis*, be confirmed in the kingdom of the flowers. I have referred above to my humble experiments in the wonderful evolutionary efforts

The Intelligence of the Flowers

of the numerous varieties of the Sage. And a curious study by Babinet on the cereals tells us that certain plants, when transported far from their habitual climate, observe the new circumstances and avail themselves of them, exactly as the bees do. Thus, in the hottest regions of Asia, Africa and America, where the winter does not kill it annually, our corn becomes again what it must have been at first, a perennial plant, like grass. It remains always green, multiplies by the root and no longer bears ears or grains. When, therefore, from its original and tropical country, it came to be acclimatized in our icy regions, it must have had to upset its habits and invent a new method of multiplication. As Babinet well says:

"The organism of the plant, thanks to an inconceivable miracle, seemed to foresee the need of passing through the grain state, lest it should perish completely during the severe season."

Life and Flowers

24

In any case, to destroy the objection which
we mentioned above and which has caused
us to travel so far from our immediate
subject, it would be enough to establish one
act of intelligent progress, were it but for
a single occasion, outside mankind. But,
apart from the pleasure which one takes
in refuting an over-vain and out-of-date
argument, how little importance, when all
is said, attaches to this question of the
personal intelligence of the flowers, the
insects or the birds! Suppose that we say,
speaking of the Orchid and the bee alike,
that it is nature and not the plant or the
insect that calculates, that combines, that
adorns, invents and thinks: what interest
can this distinction have for us? A much
loftier question and one much worthier of our
eager attention towers over these details.
What we have to do is to grasp the
character, the quality, the habits and perhaps the object of the general intelligence
whence emanate all the intelligent acts
performed upon this earth. It is from this

point of view that the study of those creatures
—the ants and the bees, among others—in
which, outside the human form, the pro-
ceedings and the ideal of that genius are
most clearly manifested becomes one of the
most curious that we can undertake. It is
clear, after all that we have shown, that
those tendencies, those intellectual methods
must be at least as complex, as advanced,
as startling in the Orchids as in the gregari-
ous *Hymenopteræ*. Let us add that a large
number of the motives and a portion of the
logic of these restless insects, so difficult of
observation, still escape us, whereas we can
grasp with ease all the silent motives, all
the wise and stable arguments of the peaceful flower.

25

Now what do we observe, when we perceive nature (or the general intelligence or
the universal genius: the name matters but
little) at work in the Orchid world ? Many
things ; and, to mention it only in passing,
for the subject would offer facilities for a
long study, we begin by ascertaining that her

idea of beauty, of gladness, her methods of attraction, her æsthetic tastes are very near akin to our own. But no doubt it would be more correct to state that ours are congenial with hers. It is, in fact, very uncertain whether we have ever invented a beauty peculiar to ourselves. All our architectural, all our musical motives, all our harmonies of colour and light are borrowed directly from nature. Without calling upon the sea, the mountains, the skies, the night, the twilight, what might one not say, for instance, of the beauty of the trees? I speak not only of the tree considered in the forest, where it is one of the powers of the earth, perhaps the chief source of our instincts, of our sense of the universe, but of the tree in itself, the solitary tree, whose green old age is laden with a thousand seasons. Among those impressions which, without our knowing it, form the limpid hollow and perhaps the subsoil of happiness and calm of our whole existence, which of us does not preserve the recollection of a few fine trees? When a man has passed mid-life, when he has come to the end of the wondering period, when he has exhausted nigh all the sights that the art, the genius and the luxury of ages and men can offer, after experiencing and comparing many things he returns to very simple memories. They raise upon the purified horizon two or three innocent, invariable and refreshing images, which he would wish to carry away with him in his last sleep, if it be true that an image can pass the threshold that separates our two worlds. For myself, I can imagine no paradise nor after-life, however splendid it may be, in which a certain magnificent Oak would be out of place, or a certain Cypress, or a Parasol Pine of Florence or of a charming hermitage near my own house, any one of which affords to the passer-by a model of all the great movements of necessary resistance, of peaceful courage, of soaring, of gravity, of silent victory and of perseverance.

26

But I am wandering too far afield : I intended only to remark, with reference to the flower, that nature, when she wishes to

be beautiful, to please, to delight and to
prove herself happy, does almost what we
should do had we her treasures at our disposal. I know that, speaking thus, I am
speaking a little like the bishop who was
astonished that Providence always made the
great streams flow close to the big cities;
but it is difficult to look upon these things
from any other than the human point of
view. Let us, then, from this point of
view, consider that we should know very
few signs or expressions of happiness if we
did not know the flower. In order well
to judge of its power of gladness and
beauty, one must live in a part of the
country where it reigns undivided, such as
the corner of Provence, between the Siagne
and the Loup, in which I am writing these
lines. Here, truly, the flower is the sole
sovereign of the hills and valleys. The
peasants have lost the habit of cultivating
corn, as though they had now only to
provide for the needs of a subtle race of
mankind that lived on sweet fragrance and
ambrosia. The fields form one great nose-
gay, which is incessantly renewed, and the

perfumes that succeed one another seem to dance their rounds all through the azure year. Anemones, Gilliflowers, Mimosas, Violets, Pinks, Narcissuses, Hyacinths, Jonquils, Mignonette, Jasmine invade the days, the nights, the winter, summer, spring and autumn months. But the magnificent hour belongs to the Roses of May. Then, as far as the eye can see, from the slope of the hills to the hollow of the plains, between dikes of Vines and Olive-trees, they flow on every side like a stream of petals whence emerge the houses and the trees, a stream of the colour which we assign to youth, health and joy. The aroma, at once warm and refreshing, but above all spacious, that opens up the sky emanates, one would think, directly from the sources of beatitude. The roads, the paths are carved in the pulp of the flower, in the very substance of Eden. For the first time in one's life, one seems to have a satisfying vision of happiness

27

Still from our human point of view and persevering in the necessary illusion, let us

add to our first remark one a little more
extensive, a little less hazardous and perhaps
big with consequences, namely, that the
genius of the earth, which is probably
that of the whole world, acts, in the vital
struggle, exactly as a man would act. It
employs the same methods, the same logic.
It attains its aim by the same means that
we would use: it gropes, it hesitates, it
corrects itself time after time; it adds,
it suppresses, it recognizes and repairs its
errors, as we should do in its place. It
makes great efforts, it invents with difficulty
and little by little, after the manner of the
workmen and engineers in our workshops.
It fights like ourselves against the heavy,
huge and obscure mass of its being. It
knows no more than we do whither it is
going; it seeks and finds itself gradually.
It has an ideal that is often confused,
but one in which, nevertheless, we distinguish a host of great lines that rise
towards a more ardent, complex, nervous
and spiritual form of existence. Materially,
it disposes of infinite resources, it knows
the secret of prodigious forces of which we

know nothing ; but, intellectually, it appears strictly to occupy our sphere : we cannot prove that, hitherto, it has exceeded its limits ; and, if it does not endeavour to take anything from beyond that sphere, does this not mean that there is nothing beyond it ? Does it not mean that the methods of the human mind are the only possible methods, that man has not erred, that he is neither an exception nor a monster, but the being through whom pass, in whom are most intensely manifested the great volitions, the great desires of the universe ?

28

The touchstones of our consciousness emerge slowly, grudgingly. Perhaps Plato's famous figure is no longer sufficient : I mean the cave with the wall above it whence the shadows of unknown men and objects are thrown into the cave below ; but, if we tried to substitute a new and more exact image in its place, this would be hardly more consoling. Suppose Plato's cave enlarged. No ray of brightness ever enters it. With the exception of light and fire, it has been carefully supplied with all that our civilization permits; and men have been imprisoned in it from their birth. They would not regret the light, having never seen it; they would not be blind, their eyes would not be dead, but, having nothing to look at, would probably become the most sensitive organ of touch.

In order to recognize ourselves in their actions, let us picture these wretches in their darkness, in the midst of the multitude of unknown objects that surround them. What quaint mistakes, what incredible deviations, what astounding misinterpretations must needs occur! But how touching and often how ingenious would seem the use which they would make of things that had not been created for employment in the dark! How often would they guess aright? And how great would not be their stupefaction if, suddenly, by the light of day, they discovered the nature and the real object of utensils and furniture which they had accommodated as best they could to the uncertainties of the shade?

The Intelligence of the Flowers

And yet their position seems simple and easy compared with our own. The mystery in which they crawl is limited. They are deprived of only one sense, whereas it is impossible to estimate the number of those in which we are lacking. The cause of their mistakes is one alone, whereas those of ours are countless.

Since we live in a cave of this sort, is it not interesting to prove that the power which has placed us there acts often and on some important points even as we act ourselves? Here we have a glimpse of light in our subterranean cave to show us that we have not been mistaken as to the use of every object to be found therein.

29

We have long taken a rather foolish pride in thinking ourselves miraculous, unparalleled and marvellously incidental beings, probably fallen from another world, devoid of any certain ties with the rest of life and, in any case, endowed with an unusual, incomparable, monstrous faculty. It is greatly preferable to be less prodigious, for we have learnt that prodigies do not take long to disappear in the normal evolution of nature. It is much more consoling to observe that we follow the same road as the soul of this great world, that we have the same ideas, the same hopes, the same trials and—were it not for our specific dream of justice and pity—the same feelings. It is much more tranquillizing to assure ourselves that, to better our lot, to utilize the forces, the occasions, the laws of matter, we employ methods exactly similar to those which it uses to conquer, enlighten and order its unsubjected, unconscious and unruly regions, that there are no other methods, that we are in the midst of truth and that we are in our right place and at home in this universe formed of unknown substances, whose thought, however, is not impenetrable and hostile, but analogous and apposite to our own.

If nature knew everything, if she were never mistaken, if, everywhere, in all her undertakings, she showed herself, at the first onset, perfect, impeccable, infallible, if she revealed in all things an intelligence immeasurably superior to our own, then there would be cause to fear and to lose courage. We should feel ourselves the victims and the prey of an extraneous power, which we should have no hope of knowing or measuring. It is much better to be convinced that this power, at least from the intellectual point of view, is closely akin to our own. Our intelligence draws upon the same reserves as does that of nature. We belong to the same world, we are almost equals. We are associating not with inaccessible gods, but with veiled, yet fraternal volitions which it is our business to surprise and to direct.

30

It would not, I imagine, be very bold to maintain that there are not any more or less intelligent beings, but a scattered, general intelligence, a sort of universal fluid that penetrates diversely the organisms which it encounters according as they are good or bad conductors of the understanding. Man would then represent, until now, upon this

earth, the mode of life that offered the least
resistance to this fluid, which the religions
called divine. Our nerves would be the
threads along which this more subtle elec-
tricity would spread. The circumvolutions
of our brain would, in a manner, form
the induction-coil in which the force of
the current would be multiplied; but this
current would be of no other nature, would
proceed from no other source than that
which passes through the stone, the star, the
flower or the animal.

But these are mysteries which it were
somewhat idle to question, seeing that we
do not yet possess the organ that could
gather their reply. Let us be satisfied with
having observed certain manifestations of
this intelligence outside ourselves. All that
we observe within ourselves is rightly open
to suspicion: we are at once judge and
suitor and we have too great an interest in
peopling our world with magnificent illusions and hopes. But let the slightest
external indication be dear and precious to
us. Those which the flowers have just
offered us are probably quite infinitesimal

The Intelligence of the Flowers

compared with what the mountains, the sea and the stars would tell us, could we surprise the secrets of their life. Nevertheless, they allow us to presume with greater confidence that the spirit which animates all things or emanates from them is of the same essence as that which animates our bodies. If this spirit resembles us, if we thus resemble it, if all that it contains is contained also within ourselves, if it employs our methods, if it has our habits, our preoccupations, our tendencies, our desires for better things, is it illogical for us to hope all that we do hope, instinctively, invincibly, seeing that it is almost certain that it hopes the same? Is it probable, when we find scattered through life so great a sum total of intelligence, that this life should make no work of intelligence, that is to say, should not pursue an aim of happiness, of perfection, of victory over that which we call evil, death, darkness, annihilation, but which is probably only the shadow of its face or its own sleep?

PERFUMES

PERFUMES

I

AFTER speaking at some length of the intelligence of the flowers, it will seem natural that we should say a word of their soul, which is their perfume. Unfortunately, here, as in the case of the soul of man, a perfume of another sphere, where reason bathes, we have at once to do with the unknowable. We are almost entirely unacquainted with the purpose of that zone of festive and invisibly magnificent air which the corollas shed around themselves. There is, in fact, a great doubt whether it serves chiefly to attract the insects. In the first place, many among the most sweet-scented of the flowers do not admit of cross-fertilization, so that the visit of the butterfly or the bee is to them a matter of indifference or inconvenience. Next, that which attracts the insects is solely the pollen and the nectar, which, generally, have no perceptible odour. And thus we see them neglect the most deliciously perfumed flowers, such as the rose and the carnation, to besiege in crowds the flowers of the maple or the hazel-tree, whose aroma is, in a manner of speaking, null.

Let us, then, confess that we do not yet know in what respect perfumes are useful to the flower, even as we cannot tell why we ourselves perceive them. Indeed, of all our senses, that of smell is the most unexplained. It is evident that sight, hearing, touch and taste are indispensable to our animal existence. A long education alone teaches us the disinterested enjoyment of forms, colours and sounds. For that matter, our sense of smell also exercises important servile functions. It is the keeper of the air we breathe, the hygienist or chemist that watches carefully over the quality of the food offered for our consumption, any disagreeable emanation revealing the presence of suspicious or dangerous germs. But besides this practical

Perfumes

mission it has another which corresponds
with nothing at all. Perfumes are utterly
useless to the needs of our physical life.
When too violent or too lasting, they may
even become detrimental to it. Nevertheless, we possess a faculty that revels in them
and brings us the joyful tidings of them with
as much enthusiasm and conviction as though
it concerned the discovery of a delicious
fruit or drink. This uselessness deserves
our consideration. It must hide some fair
secret. We have here the only occurrence
in which nature procures us a gratuitous
pleasure, a satisfaction that does not serve to
adorn one of necessity's snares. Our scent
is the only purely luxurious sense that she
has granted us. Wherefore it seems almost
foreign to our bodies, appears to be not very
closely connected with our organism. Is it
an apparatus that is developing, or one that
is wasting away; a somnolent, or an awakening faculty? Everything leads us to think
that it is being evolved on even lines with
our civilization. The ancients interested
themselves almost exclusively in the more
brutal, the heavier, the more solid scents, so

to speak: musk, benzoin, incense; and the
fragrance of the flowers is very seldom mentioned in Greek and Latin poetry or in
Hebrew literature. To-day, do we ever see
our peasants, even at their longest periods
of leisure, dream of smelling a violet or a
rose? And is not this, on the other hand,
the very first act of an inhabitant of our
great cities who perceives a flower? There
is, therefore, some ground for admitting
that the sense of smell is the last-born of
our senses, the only one, perhaps, that is
not "in course of retrogression," to use the
ponderous phrase of the biologists. This
is a reason for making it our study, ques-
tioning it and cultivating its possibilities.
Who shall tell the surprises which it would
have in store for us if it equalled, for
instance, the perfection of our sight, as it
does in the case of the dog, which lives as
much by the nose as by the eyes?

We have here an unexplored world.
This mysterious sense, which, at first sight,
appears almost foreign to our organism,
becomes, perhaps, when more carefully considered, that which enters into it most

intimately. Are we not, above all things, beings of the air? Is the air not for us the most absolutely and promptly indispensable element; and is not our smell just the one sense that perceives some parts of it? Perfumes, which are the jewels of that life-giving air, do not adorn it without good cause. It were not surprising if this luxury which we do not understand corresponded with something very profound and very essential and rather, as we have seen, with something that is not yet than with something that has ceased to be. It is very possible that this sense, the only one that is turned towards the future, is already discerning the most striking manifestations of a form or of a happy and salutary state of matter that is reserving many surprises for us.

Meanwhile, it has not yet reached beyond the stage of the more violent, the less subtle perceptions. Hardly does it so much as suspect, with the aid of the imagination, the profound and harmonious effluvia that evidently envelop the great spectacles of the atmosphere and the light. As we are on the

point of distinguishing those of the rain and
the twilight, why should we not one day
succeed in recognizing and fixing the scent
of snow, of ice, of morning dew, of the first
fruits of the dawn, of the twinkling of the
stars; for everything must have its perfume,
inconceivable, as yet, in space: even a
moonbeam, a ripple of water, a hovering
cloud, an azure smile of the sky. . . .

2

Chance or rather the choice of life has
brought me back lately to the spot where
almost all the perfumes of Europe are born
and elaborated. It is, in point of fact, as
every one knows, in the sun-swept and
balmy region stretching from Cannes to Nice
that the last hills and the last valleys of
living and true flowers maintain an heroic
struggle against the coarse chemical odours
of Germany, which stand in exactly the
same relation to nature's perfumes as do the
painted woods and plains of a theatre to
the woods and plains of the real country.
Here the labourer's work is ruled by a

sort of purely floral calendar, in which, in
May and July, two adorable queens hold
sway: the rose and the jasmine. Around
these two sovereigns of the year, one the
colour of the dawn, the other clad in white
stars, defile in procession, from January to
December, the violets, innumerous and
prompt, the artless, marvel-eyed narcissuses,
the clustering mimosas, the mignonette, the
pink laden with precious spices, the imperious geranium, the tyrannically virginal
orange-flower, the lavender, the Spanish
broom, the too-potent tuberose and the
acacia that resembles an orange caterpillar.

It is, at first, not a little disconcerting to
see the great dull and heavy rustics, whom
harsh necessity turns every elsewhere from
the smiles of life, taking flowers so seriously,
handling carefully those fragile ornaments
of the earth, performing a task fit for a
princess or a bee and bending under a weight
of violets or jonquils. But the most striking impression is that of certain evenings or
mornings in the season of the roses or the
jasmine. It is as though the atmosphere of
the earth had suddenly changed, as though

it had made way for that of an infinitely
happy planet, where perfumes are not, as
here, fleeting, vague and precarious, but
stable, spacious, full, permanent, generous,
normal and inalienable.

3

Many writers, speaking of Grasse, have
drawn the picture of that almost fairy-like
industry which occupies the whole of a hard-
working town, perched, like a sunlit hive,
upon a mountain-side. They have told of
the magnificent cartloads of roses shot upon
the threshold of the smoking factories, the
great halls in which the sorters literally
wade through the flood of petals, the less
cumbersome, but more precious arrival of
the violets, tuberoses, acacias, jasmine, in
wide baskets which the peasant-women carry
nobly on their heads. Lastly, they have
described the different processes whereby
the flowers, each according to its character,
are forced to yield to the crystal the marvellous secrets of their hearts. We know
that some of them, the roses, for instance,

Perfumes

are accommodating and willing and give up their aroma with simplicity. They are heaped into huge boilers, tall as those of our locomotive engines, through which steam is made to pass. Little by little, their essential oil, more costly than a jelly of pearls, oozes drop by drop into a glass tube, no wider than a goose-quill, at the bottom of the monstrous still, which resembles some mountain painfully giving birth to a tear of amber.

But the greater part of the flowers do not so easily allow their souls to be imprisoned. I will not, in the wake of so many others, speak here of the infinitely varied tortures inflicted upon them to force them at length to surrender the treasure which they desperately hide in the depth of their corollas. I will not enumerate the different processes of chemical extraction by means of petrol ether, sulphide of carbon and the rest. The great perfumers of Grasse, ever faithful to tradition, scorn these artificial and almost unfair methods, which wound the soul of the flower. It will suffice, to give an idea of the executioner's cunning and the obstinacy of

some of the victims, to recall the pangs of
the enfleurage which certain flowers are made
to endure before they break silence. The
cold enfleurage is practised only upon the
jonquil, the mignonette, the tuberose and
the jasmine; [1] and I may mention, in passing,
that the scent of the jasmine is the only one
that is inimitable, the only one that cannot
be obtained by a cunning mixture of other
odours.

The torturer coats large plates of glass
with a white fat of the thickness of two
fingers and spreads on this bed of humiliating
pain the flowers to be questioned. As the
result of what hypocritical manœuvres, of
what unctuous promises does the fat obtain

[1] The violets resist the seductions of the cold fat; and
the torture of fire has to be superadded. The lard,
therefore, is heated in the water-bath until it approaches
boiling-point. In consequence of this barbarous treat-
ment, which recalls that inflicted upon the coiners of the
middle ages, the modest and fragrant flowers that deck
the roads in spring gradually lose the strength to keep
their secret. They yield, they surrender; and their
liquid executioner is not satiated until it has absorbed
four times its own weight in petals, which causes the
torture to be prolonged throughout the season in which
the violets blossom under the olive-trees.

their irrevocable confidences? None can
tell; but the fact remains that soon the
too-trusting flowers have nothing more to
lose. Forthwith, they are removed and flung
away as rubbish; and, each morning, a fresh
ingenuous heap come to take their place on
the insidious couch. These yield in their turn
and undergo the same fate; others and yet
others follow them; and it is not until the
end of three months, that is after devouring ninety successive layers of flowers, that
the unctuous ogre is completely surfeited
and refuses to absorb the life and soul of
any further victims.

It now becomes a matter of making the
wan miser disgorge; for the fat means,
with all its shapeless and evasive energy,
to retain the absorbed treasure. This is
achieved, not without difficulty. The fat
has base passions which are its undoing. It
is plied with alcohol, is intoxicated and
ends by quitting its hold. The alcohol
now possesses the mystery. No sooner has
it the secrets in its custody than it too
claims the right to impart them to none
other, to keep them for itself alone. It is

Life and Flowers

attacked in its turn, tortured, evaporated,
condensed; and, after all these adventures,
the liquid pearl, pure, essential, inexhaustible
and almost imperishable, is at last gathered
on a crystal blade.

THE END

www.ingramcontent.com/pod-product-compliance
Lightning Source LLC
Chambersburg PA
CBHW010728270326
41930CB00016B/3404